AF404729

Ion Channel Gating and Mechanisms

Online at: https://doi.org/10.1088/978-0-7503-2388-8

Committee Chairperson

Les Satin
University of Michigan, USA

Editorial Advisory Board Members

Geoffrey Winston Abbott
UC Irvine, USA

Da-Neng Wang
New York University, USA

Mibel Aguilar
Monash University, Australia

Kathleen Hall
Washington University in St Louis, USA

Cynthia Czajkowski
University of Wisconsin, USA

David Sept
University of Michigan, USA

Miriam Goodman
Stanford University, USA

Andrea Meredith
University of Maryland, USA

Jim Sellers
NIH, USA

Leslie M Loew
University of Connecticut School of Medicine, USA

Joe Howard
Yale University, USA

Meyer Jackson
University of Wisconsin, USA

About the Series

The Biophysical Society and IOP Publishing have forged a new publishing partnership in biophysics, bringing the world-leading expertise and domain knowledge of the Biophysical Society into the rapidly developing IOP ebooks program.

The program publishes textbooks, monographs, reviews, and handbooks covering all areas of biophysics research, applications, education, methods, computational tools, and techniques. Subjects of the collection will include: bioenergetics; bioengineering; biological fluorescence; biopolymers *in vivo*; cryo-electron microscopy; exocytosis and endocytosis; intrinsically disordered proteins; mechanobiology; membrane biophysics; membrane structure and assembly; molecular biophysics; motility and cytoskeleton; nanoscale biophysics; and permeation and transport.

A list of recently published title in this series can be found here: https://iopscience. iop.org/bookListInfo/iop-series-in-biophysical-society.

Ion Channel Gating and Mechanisms

David T Yue
Department of Biomedical Engineering and Neuroscience, Johns Hopkins University
Deceased

Manu Ben-Johny
Department of Physiology and Cellular Biophysics, Columbia University,
New York, NY, USA

Ivy E Dick
Department of Pharmacology and Physiology, University of Maryland-Baltimore,
Baltimore, MD, USA

IOP Publishing, Bristol, UK

ISBN   978-0-7503-2388-8 (ebook)
ISBN   978-0-7503-2386-4 (print)
ISBN   978-0-7503-2389-5 (myPrint)
ISBN   978-0-7503-2387-1 (mobi)

DOI   10.1088/978-0-7503-2388-8

Version: 20260101

IOP ebooks

British Library Cataloguing-in-Publication Data: A catalogue record for this book is available from the British Library.

Published by IOP Publishing, wholly owned by The Institute of Physics, London

IOP Publishing, No.2 The Distillery, Glassfields, Avon Street, Bristol, BS2 0GR, UK

US Office: IOP Publishing, Inc., 190 North Independence Mall West, Suite 601, Philadelphia, PA 19106, USA

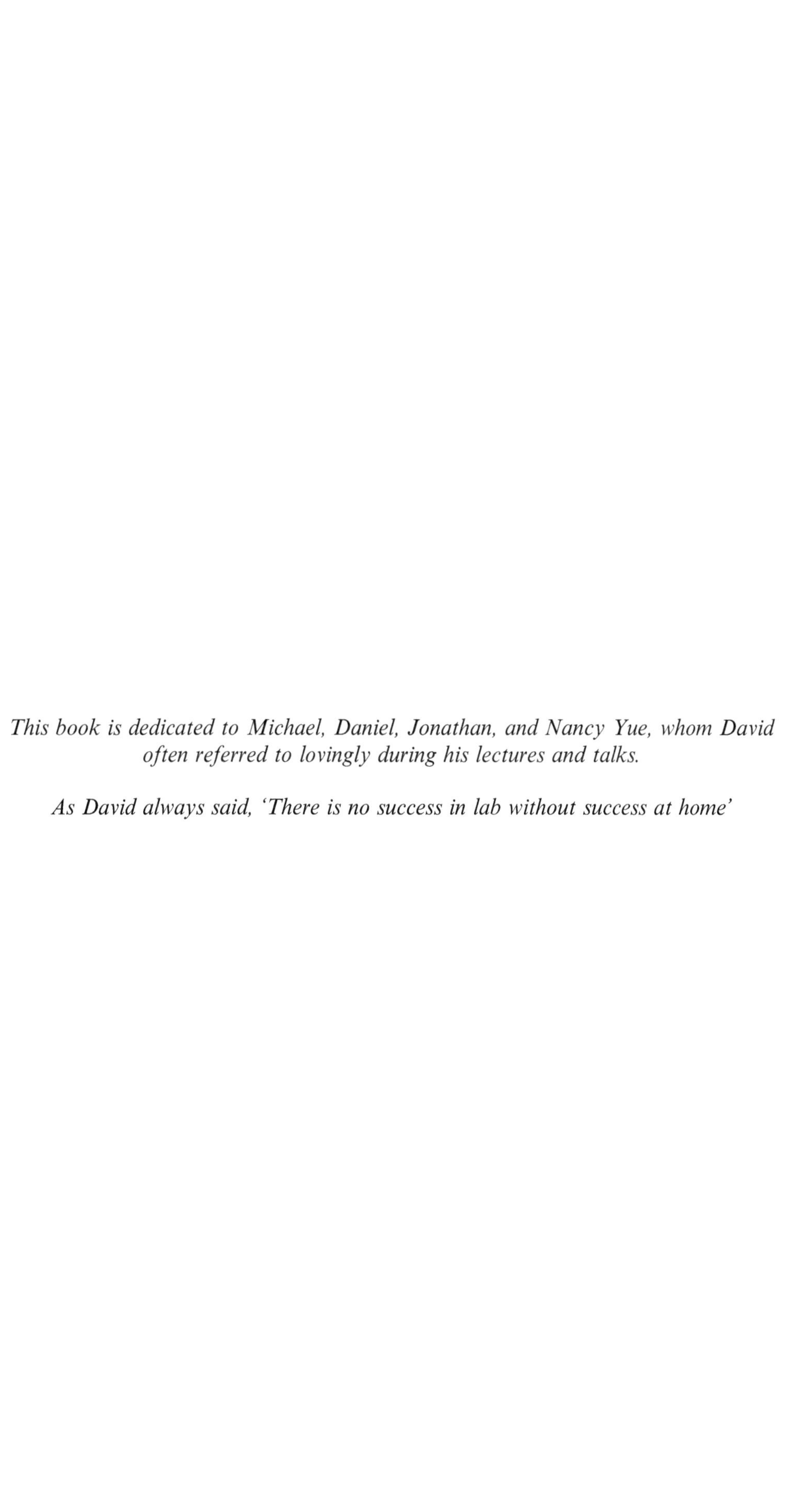

This book is dedicated to Michael, Daniel, Jonathan, and Nancy Yue, whom David often referred to lovingly during his lectures and talks.

As David always said, 'There is no success in lab without success at home'

Contents

5 Fluctuation analysis of ion-channel gating

6 Gating currents: an alternate view of channel operation

Preface

This book is based on the writings and teachings of the late Dr David Tuckchow Yue (1957–2014), professor in the Department of Biomedical Engineering at Johns Hopkins University and a renowned biophysicist. Through his passion for science and his distinctive teaching style, David inspired numerous undergraduate and graduate students, along with postdoctoral fellows, leaving a lasting imprint on the field of ion-channel biophysics.

This book is organized into six chapters, each originally written by Dr David Yue. The chapters were compiled and reformatted, with the addition of a few illustrations, to be in book format. These chapters formed foundational material for a popular course that Dr Yue taught at Hopkins entitled *Ion Channels of Excitable Membranes*. The first chapter, which reviews the classical view of ion channels through the lens of Hodgkin and Huxley, was also part of an undergraduate course entitled *Physiological Foundations*, later known as *Systems Bioengineering I*. We have strived to remain as faithful as possible to the original version written by David.

The material for these chapters draws heavily from several seminal publications in the field and from the classic book *Ion Channels of Excitable Membranes* by Dr Bertil Hille. These were companion materials for the course. While additional material based on current topics was often included in the class, this content varied year by year and was not accompanied by formal text written by David; it is thus not included in this book. Instead, we focus on classic concepts that pertain to ion-channel function/gating, the central focus of David's course.

David Yue was a gifted teacher who is remembered for his eloquence, clarity, and enthusiasm. Despite the challenging nature of the course material, his lectures were adored by his students. David devoted significant time to preparing and perfecting his lectures. He combined humor and deep thinking in a manner that was both enjoyable and engaging—a style unique to himself. One of the devices he used effectively in his teaching was analogies or 'scientific parables' to break down difficult concepts in a manner that was relatable and certainly memorable. We have included several of these analogies throughout the chapters. These are entirely in his words, as transcribed from recordings of his lectures by his trainees.

A key emphasis in his courses and in his laboratory (the Calcium Signals Lab) was the importance of being precise and quantitative. He often invited trainees to the whiteboard to articulate their thoughts and ideas, handing them a marker and gently challenging them: 'Say what you mean and mean what you say.' Although at times intimidating, the act of laying out step by step the logic of a physiological argument and working through ad hoc derivations was invaluable training for his students to gain a deep understanding of the material and to become effective communicators. In like manner, David often emphasized that if you truly understand a physiological phenomenon, you should be able to write an equation to describe it. This combination of rigor and thoughtfulness deeply influenced his students and trainees, who are now spread across the world.

David was a brilliant scientist and an inspirational leader in the field of ion channels. His passion for science, his sense of wonder, and his unrelenting pursuit of truth are deeply missed. David continues to inspire us in our approach to science and our approach to teaching. By compiling these lectures now, ten years after his passing, we hope that David's lectures will continue to inspire the next generation of scientists and thinkers.

Ivy E Dick, PhD
Associate Professor
Department of Pharmacology & Physiology
University of Maryland Baltimore

Manu Ben-Johny, PhD
Assistant Professor
Department of Physiology and Cellular Biophysics
Columbia University

Every so often, the veil of confusing experimental results is parted, and something deep and beautiful about how biological life works is revealed. It is as if a syllable that God spoke becomes suddenly audible. The thrill of unearthing such 'God speak' is one of the special rewards of my profession.
—David T Yue

Acknowledgments

We thank the members of the Calcium Signals Lab at Johns Hopkins University throughout the years who were deeply inspired by David Yue, and who contributed to development of this material.

Thank you to: Paul Adams, Heather Agler, Badr Alseikhan, Rebecca Alvania, Molly Anderson, Edwin Antony, Jennifer Babich, Rahul Banerjee, Hojjat Bazzazi, Peter Bedner, Manu Ben-Johny, David Brody, Dongming Cai, Florence Chan, Dipayan Chaudhuri, Myoung Hyun (Brian) Choi, Henry M. Colecraft, Stanislava Dalton, Marita de Leon, Carla DeMaria, Ivy Dick, Lynn Dobrunz, Johnny Elfar, Mike Erickson, Jenafer Evans, Hadi Fatemi-Shariatpanahi, Kevin Gingrich, Wei Guo, Stefan Herzig, John Imredy, John Issa, Lisa Jones, Rosy Joshi-Mukhurjee, Po Wei Kang, Timothy Kernan, Joanna Lee, Shin Rong Lee, Bert Lewis, Haoya Liang, Jiangyu Li, Sean Lie, Xiaodong Liu, Worawan Limpitikul, Jayalakshmi Miriyala, Jennifer Mulle, Tracy Morrison, Daniel Moon, Manning Zhang, Karl Zhang, Masayuki Mori, Joel Musee, Ho Namkumg, Lan Le Ngoc, Trang Nguyen, Jacqueline Niu, Nick Perchiniak, Blaise Peterson, Parag Patil, Sarah Park, Lingjie Sang, Michael Shen, Jamie Stratton, Michael Tadross, Shoji Takahashi, Lai Hock Tay, Balaji Veeramani, Yan Wang, Shao-Kui Wei, Phil Yang, and Wanjun Yang.

We are proud to have been trained by David Yue.

Author biographies

David Yue

David T Yue, M.D., Ph.D., was a renowned biophysicist and Professor of Biomedical Engineering and Neuroscience at Johns Hopkins University. He graduated *magna cum laude* from Harvard University in 1979 with a degree in Biochemistry, and later earned both his Ph.D. in Biomedical Engineering and his M.D. from Johns Hopkins in 1987. Dr. Yue joined the Department of Biomedical Engineering at Johns Hopkins in 1988, where he spent the entirety of his distinguished career. Widely regarded as a brilliantly creative and intellectually rigorous scientist, Dr. Yue made foundational contributions to our understanding of ion channel biophysics. His achievements were recognized with numerous honors, including the prestigious Kenneth S. Cole Award from the Biophysical Society. He was also an extraordinary mentor and a beloved teacher who inspired countless undergraduate and graduate students, as well as postdoctoral fellows. Dr. Yue passed away tragically on December 23, 2014, leaving behind a lasting scientific legacy and a deeply felt personal impact on all who knew him.

Manu Ben-Johny

Manu Ben-Johny, Ph.D. is an Assistant Professor in the Department of Physiology and Cellular Biophysics at Columbia University. He received a BS in Biomedical engineering and Mathematics from Saint Louis University and completed his PhD in Biomedical Engineering at Johns Hopkins University under the tutelage of Dr. David Yue. His lab focuses on understanding ion channel dysfunction in neurological and cardiac diseases, with a focus on engineering custom modulators that restore proper ion channel function in disease.

Ivy E Dick

Ivy Dick, Ph.D. is an Associate Professor in the Pharmacology and Physiology Department at the University of Maryland, School of Medicine. She began her career under the direction of Drs. Charles Cohen and Owen McManus at Merck Research Labs and subsequently earned her doctoral degree under the direction of Dr. David Yue in the Biomedical Engineering department at Johns Hopkins University, where she studied the mechanisms underlying calmodulin regulation of voltage-gated calcium channels. Dr. Dick's current research focuses on understanding the mechanisms of voltage-gated calcium channel regulation, and how genetic mutations disrupt channel function.

IOP Publishing

Ion Channel Gating and Mechanisms

David T Yue, Manu Ben-Johny and Ivy E Dick

Chapter 1

Hodgkin–Huxley view of ion channels

1.1 Introduction

The purpose of this section is to review the elegant, functional way in which Hodgkin and Huxley (HH) described the function of ion channels. In five classic papers published in 1952 (Hodgkin and Huxley 1952d, 1952c, 1952a, 1952b, Hodgkin *et al* 1952), they elaborated a quantitative viewpoint for voltage-gated Na^+ and K^+ channels. This conceptualization was so far advanced that the basic features of their concept are still valid and useful as a starting point. In particular, they split the performance characteristics of ionic channels into two broad classes: permeation and gating. Permeation describes how ions pass through the channel when it is in an 'open' or 'conducting' conformation. Gating describes how the channel opens and closes (Hille 2001).

1.2 Commonsense approximations of Hodgkin and Huxley

1.2.1 Ion channels are not always on; transport is gated

HH found that the response of nervous tissue to step changes in voltage was not linear. Negative-going steps from the resting potential resulted in passive 'RC' decays of current. Conversely, positive-going steps exceeding a threshold resulted in action potentials, clearly demonstrating nonlinear behavior. If ion-channel transport properties were time-invariant, nervous tissue should behave something like a network of capacitors and resistors. Hence, they suspected that ion channels were gated, with gates that could sense the transmembrane potential (V_m).

1.2.2 Cell membrane is like a capacitor

The cell membrane in which ion channels are situated comprises a phospholipid bilayer, as shown in figure 1.1. The hydrophobic tails of the phospholipids form a

doi:10.1088/978-0-7503-2388-8ch1 1-1

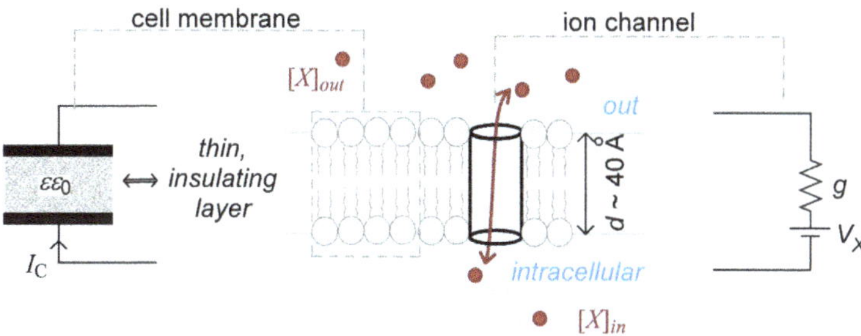

Figure 1.1. Schematic showing the cell membrane and the open pore of an ion channel. The cell membrane can be viewed as a capacitor, while the open ion channel can be viewed as a resistor.

thin insulating layer, while ions may accumulate near the hydrophilic or charged lipid head groups. Hence, the cell membrane can be viewed as a capacitor and thereby has the following properties:

$$C \propto \frac{\varepsilon\varepsilon_0 \times A}{d},$$

where ε is the dielectric constant, ε_0 is the polarizability of free space, A represents the surface area of the membrane, and d is the thickness of the bilayer.

From basic electric field theory, C has the following operational features:

$$C = \frac{Q}{V_m} \Rightarrow I_C = C\frac{dV_m}{dt},$$

where Q represents the charge accumulated at the membrane. The current $I > 0$ when positive ions flow from inside to outside, according to our sign convention for the transmembrane potential, $V_m = \psi_{in} - \psi_{out}$.

1.2.3 A nonselective channel pore behaves according to Ohm's law

If the open pore of an ion channel is viewed as a simple resistor, then the following equation holds:

$$I(\text{amp}) = g(\text{siemen} = \Omega^{-1}) \times V_m(\text{volt}),$$

where g represents the conductance.

For orientation, it is useful to recall that an ampere is a coulomb per second; an elementary charge $e = 1.6 \times 10^{-19}$ coulombs; and there are Avogadro's number ($N_A = 6.022 \times 10^{23}$) of molecules in a mole. Faraday's constant (F) is the charge equal to a mole of elementary charges:

$$F = N_A e \approx 10^5 \,(\text{coulombs/mole}).$$

Another useful entity is RT/F, where R is the gas constant and T is the temperature in Kelvin. $RT/F \sim 25$ mV at room temperature.

1.2.4 Ion channels offer selective passage of one type of ion, requiring a battery offset

Earlier indirect experiments suggested that there were different transport processes that were selective for Na^+ on the one hand and K^+ on the other. If we assume perfect selectivity, then only one type of ion can pass through the open pore of a channel. It turns out that Na and K channels are quite selective, which is the reason we call them 'Na' and 'K' channels. Given this high selectivity, a simple resistor alone does not make sense as a description of the ion-channel pore. Consider a Na channel, with $[Na]_{out} \gg [Na]_{inside}$. Intuition tells us that if $V_m = 0$, then $I_{Na} = g \times V_m = 0$ would not be true. Instead, the Boltzmann distribution, taken from statistical thermodynamics, tells us what the transmembrane voltage V_m must be when $I = 0$. The Boltzmann distribution specifies the ratio of the probabilities of finding a molecule in state i versus state j when the system is at thermodynamic equilibrium:

$$\frac{P_i}{P_j} = \exp\left(-\frac{U_i - U_j}{RT}\right),$$

where U_i and U_j are the energy changes associated with introducing a mole of molecules into state i or j, respectively. A specific application is to specify the relative equilibrium probabilities of finding an ion X on the inside or outside of the membrane, based on the relative energy of an ion residing on the inside or outside of the membrane. Hence,

$$\frac{[X]_{out}}{[X]_{in}} = \exp\left(-\frac{U_{out} - U_{in}}{RT}\right).$$

Rearranging this equation and recalling that the electrical work required to move a mole of charge against an electric field is $V \times z_X \times F$ yields (see Box 1.1):

$$RT \times \ln\left(\frac{[X]_{out}}{[X]_{in}}\right) = U_{in} - U_{out} = V_X \times z_X \times F.$$

Box 1.1. Population control through economics—an analogy for the Nernst potential.

Imagine you could teleport anywhere in the world. Where would you go? Perhaps somewhere with less people like Hawaii! You could work on your tan, make sand castles, surf, snorkel, hike, etc. If people could teleport, eventually at equilibrium, the population density in Hawaii would probably equal the population concentration here in Baltimore. Thus, in the absence of an electric field the concentration of ions will equal on both sides of the membrane. Unfortunately, teleportation is not yet feasible so if you want to go to Hawaii you have to take a plane, which is extremely expensive. Thus, although the population density in Hawaii is less than that in Baltimore, travel to Hawaii is limited by economics such as it might be for ions in the presence of an electric field.

—David T. Yue

Because this equation only applies to X at equilibrium, which, in this case, corresponds to $I = 0$, V_X is the transmembrane potential at which zero current flows. V_X is known as the Nernst potential for ion X. Hence, thermodynamics gives us one concrete constraint about conduction through a selective ion channel: $I = 0$ at $V_m = V_X$:

$$V_X = \frac{RT}{z_X \times F} \times \ln\left(\frac{[X]_{\text{out}}}{[X]_{\text{in}}}\right).$$

What about the case where V_m is not near V_X? Over a limited voltage range, we might suspect that the I–V relation for the open channel would be approximately linear, a suspicion that HH were able to verify to a reasonable extent. Hence,

$$I = g \times (V_m - V_X),$$

which is equivalent to the battery–resistor circuit in figure 1.1.

One interpretation of this approximation follows when we express V_X explicitly in terms of concentrations, as above:

$$I = g \times V + g \times \frac{RT}{z_X \times F} \times \ln\left([X]_{\text{in}}\right) - g \times \frac{RT}{z_X \times F} \times \ln\left([X]_{\text{out}}\right).$$

Table 1.1 summarizes common Nernst potentials for various ions in cells (Hille 2001). These would correspond to the values of various battery offsets for the different types of ion channels (see Box 1.2).

Table 1.1. Free ionic concentrations and equilibrium potentials for skeletal muscle.

Ion	Extracellular concentration (mM)	Intracellular concentration (mM)	Equilibrium potential (mV)
Na^+	145	12	+67
K^+	4	155	−98
Ca^{2+}	1.5	$<10^{-7}$ M	+128
Cl^-	123	4.2	−90

Box 1.2. Bruce Lee and the salty wound—an 'mnemonic' for ion concentrations in cells.

Remember what happens to Bruce Lee when he gets into a fight. Bruce lee is probably the greatest Kung Fu artist that has ever lived. It is incredible. He fought with a lot of wicked villains. One that I remember is Mr. Han man. If you remember Bruce Lee, the fights didn't always go smoothly. At the end victory would prevail but initially it was kind of rough. Usually, he would come in there and there would be some wicked villain, and Bruce Lee wouldn't be quite ready and the wicked villain of course would strike first. He was struck and it was kind of unfair, because the wicked villain didn't fight with human hands. He had an attachment with sharp blades on it. So, the villain would scratch Bruce Lee. Bruce Lee wasn't fast enough in the initial phase to get out of the way. So, there is

> *blood, and what do you do in this sort of thing here—tasted that blood and that got him going. When he tasted that blood what did he taste? What was it like? Salty! Haven't you ever been punched in the face? It is salty. The reason it is salty is because your blood is like the extracellular solution of cells and there is a lot of NaCl. So, there is high sodium concentration. And on the inside, it is lower. What about potassium, potassium is kind of the opposite.*
>
> —David T. Yue

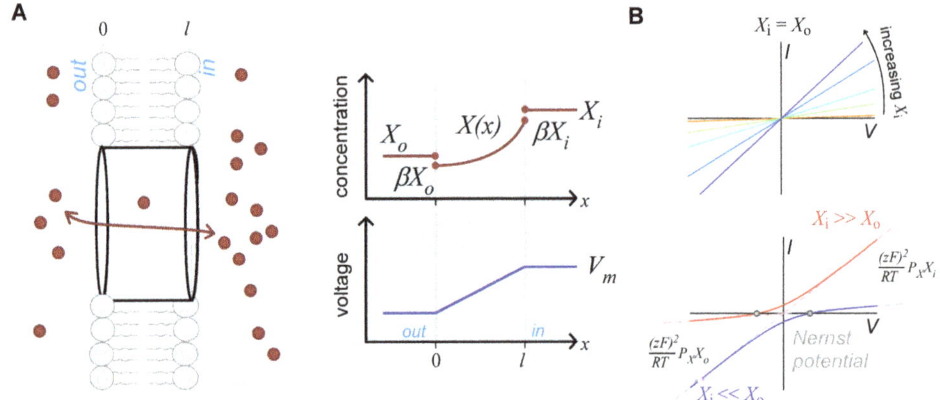

Figure 1.2. (A) Layout of the GHK current equation model. Ions within the channel pore are assumed to be independent, and the electric field across the membrane is assumed to be constant. (B) Schematic of the components of ion flux: diffusion and drift due to the electric field.

Implicit in the assumption of selectivity is the notion that there are separate types of channels for the transport of the different types of ions.

1.2.5 Electrodiffusion model—the Goldman–Hodgkin–Katz equation

One of the earliest attempts to address these problems was made by Hodgkin *et al* using the Nernst–Planck electrodiffusion equation (Hodgkin and Katz 1949). We develop this in detail because so much of our 'first-order approximation' logic derives from this model (Goldman 1943).

In this model, the channel pore is viewed as an isotropic slab of material. The layout is as shown in figure 1.2. This layout can be solved for the GHK current equation as follows. Start with the Nernst–Planck electrodiffusion equation, which describes current flow through this slab of material (see Box 1.3):

$$I_{x,\,\text{in}\to\text{out}} = +z_X \times F \times D_X \left(\overbrace{\frac{dX(x)}{dx}}^{\text{part a}} + \overbrace{X(x) \times \left(\frac{z_X F}{RT} \times \frac{d\psi}{dx} \right)}^{\text{part b}} \right).$$

Box 1.3. Mom's home—an analogy for the Goldman–Hodgkin–Katz equation.

...in the absence of electric field you are just bouncing—random walks that is the Fick's law part. This is like my boys and I before mom comes home. Often times we are just chilling out in the family room and it is a little bit messy and we got chips out, soda out. Dad doesn't care too much. We are just diffusing—random walk. But when the electric field is turned on right, that is like when mom is getting home. All of a sudden, we feel something. 'Oh, it is not cleaned up, and there is a certain thing.'... So if the electric field is on you will feel a force ... So you move, but when you start moving, it is like Mom says 'Dad, clean up the room—clean up this, that. But of course you can't clean things up instantaneously—you would run into things in the room and slow down. So there is a friction coefficient. Right? So eventually the electrical field is countered by the frictional force and you reach sort of a terminal velocity. There is only so fast that you can clean up the house, much to Mom's concern.'

—David T. Yue

Part **a** of the Nernst–Planck electrodiffusion equation has to do with simple diffusion, tantamount to Fick's law or random walks. Part **b** involves the establishment of a certain drift velocity determined by the balance of friction and electrical force applied to a permeant ion in the channel.

To understand part **b**, consider ions residing in the pore of the channel. The transmembrane voltage gradient imposes an electrical force on these ions, which is counterbalanced by a frictional force that is proportional to their drift velocity ($\langle v \rangle$):

$$0 = -f_x \langle v \rangle - Z_x F \frac{d\psi}{d\chi} \text{ (net force is thereby zero).}$$

Here, f_x is a proportionality constant that is inversely proportional to mobility. Therefore,

$$\langle v \rangle = \frac{-Z_x F}{f_x} \frac{d\psi}{d\chi}.$$

As mobility is proportional to the diffusion coefficient,

$$\frac{1}{f_x} = \mu_x = \text{mobility} = \frac{D_x}{RT} \rightarrow \langle v \rangle = \frac{-Z_x D_x F}{RT} \frac{d\psi}{d\chi}.$$

The ion flux is measured in moles of ions per unit time, per unit area. It is negative because of the convention used for direction, namely $i \rightarrow o$:

$$J_{x,\,1-0} = -X(\chi)\langle v \rangle = +X(\chi)\frac{z_x D_x F}{RT} \frac{d\Psi}{d\chi}.$$

Current (charge per unit time, per unit area) is then given by:

$$I_{x,\,1-0} = z_x F J_{x,\,1+0} = +X(\chi)\frac{D_x}{RT}(z_x F)^2 \frac{d\psi}{d\chi}.$$

The addition of this term to Fick's law yields the Nernst–Planck electrodiffusion equation above.

The solution to this equation, assuming a constant electric field, is:

$$I_{x,\ \text{in}\rightarrow\text{out}} = \frac{(z_X F)^2}{RT} \times \frac{D_X \beta}{l} \times V_m \left(\frac{X_{\text{in}} - X_{\text{out}} \times \exp\left(-\frac{z_X F}{RT} V_m\right)}{1 - \exp\left(-\frac{z_X F}{RT} V_m\right)} \right).$$

Substituting the 'permeability coefficient for ion X,' defined as the entity $P_X = D_X \beta / l$, yields the final equation. This equation is the famous Goldman–Hodgkin–Katz (GHK) current equation:

$$I_{x,\ \text{in}\rightarrow\text{out}} = \frac{(Z_X F)^2}{RT} \times P_X \times V_m \left(\frac{X_{\text{in}} - X_{\text{out}} \times \exp\left(-\frac{Z_X F}{RT} V_m\right)}{1 - \exp\left(-\frac{Z_X F}{RT} V_m\right)} \right).$$

It is worth learning how to sketch the predicted behavior of the GHK current equation, as illustrated in Figure 1.2(B–C). Note the use of extreme-voltage asymptotes to produce the sketch.

Two important features of ion channels become readily apparent from this analysis. First, the conductance of an ion channel varies with different concentrations of permeant ions (see Box 1.4). This is intuitive, as water without any ions does not

---

Box 1.4. The bathtub and the inconsiderate roommate—an analogy for ion permeation.

Suppose you have your roommate from hell that you are forced to live with. And you are very good citizen at the beginning of the year. So you put up with them. But this roommate is really bad and by the time a few months in the semester, it is getting old. You have these very evil thought. So you want to get rid of this roommate but you don't know how. So one day, one of the things that bothers you about this evil roommate is that they are hogging the bathroom in the morning—nothing worse than that—right? Not only are they hogging it, they are taking a nice long bath. Who takes a bath in the morning? So you can't get ready for the day and you can't make it to SBE. And this bad roommate is not only bad to you but he is bad to others. He works in a lab and he is taking a bath using water stolen from his lab—the double distilled water—highest quality water. So you don't know that detail at first so you say, 'are you almost done with the bath tub?' He responds, 'Stop rushing me!' So you see this little radio by there with a power plug and you knock it into the bathtub and you are thinking you will electrocute this roommate and you will be done with it. So you knock it in there, and it splashes. Nothing happens! This evil roommate says, 'hey what are you doing?' And suddenly you remember the lecture from Systems Bioengineering I about permeation. With the double-distilled water there are hardly any ions in it. The conductivity is very low and you can't electrocute them. What I need is some salt in there. So you say, 'Just a moment.' You go back out to the kitchen and you get the salt shaker, you unscrew it and put a little a handful of white crystalline salt in there. And you come back in there, and you just flick it out there and as the salt is coming in you say, 'hasta la vista, babe!' He is electrocuted. So you will never forget it again.

—David T. Yue

conduct electricity, and the conductance increases as ion concentration increases. For real ion channels, the conductance saturates as the permeant ion concentration increases. The GHK model does not account for this. Considering ion binding to the pore could provide an explanation for the saturation of ion conductance. Second, when the concentrations of a permeant ion are unequal on the inside and outside, the amplitudes of inward versus outward currents are different—a phenomenon called rectification.

With the current equation in hand, we can proceed to engage another famous perspective of this model, as embodied in the GHK voltage equation. The heart of this latter equation comes from the realization that we can now predict I_T, even when the channel is not perfectly selective. Using the property of independence, we can write:

$$I_T = I_1 + I_2 + I_3 + \ldots + I_n,$$

where each of the I_i terms on the right is the GHK current equation for one of the different types of permeant ions in an imperfectly selective channel.

To predict the reversal potential (V_{rev}), set $I_T = 0$, yielding

$$I_T = 0 = I_{Na} + I_K + I_{Cl},$$

where Na, K, and Cl may be permeant ions in the generic channel in question. Plugging in explicit GHK current equations for the Na, K, and Cl terms and solving yields the GHK voltage equation stated below:

$$V_{rev} = -\frac{RT}{F} \ln \left(\frac{P_{Na} \times Na_{in} + P_K \times K_{in} + P_{Cl} \times Cl_{out}}{P_{Na} \times Na_{out} + P_K \times K_{out} + P_{Cl} \times Cl_{in}} \right).$$

1.3 Hodgkin–Huxley preconception

Using these commonsense assumptions/approximations, HH already had a firm suspicion about what they would see in their voltage-clamp experiments. The assumptions above could be brought together in the following model of cell membrane excitability (see Box 1.5).

Box 1.5. A Christmas gift for the boys—analogy for ion channels as a circuit model.

So, when I was a younger professor, I had three young boys you know they were all below ten, were ten years and diapers, running around the house. A lot of times, they wasted their childhood in my sort of viewpoint—electronic video games and card games and things like that. And sort of as a Hopkins professor coming from a nerd world view, I wanted to interest them in the higher things in life. So, when it came to Christmas, I wanted to give them a gift that not only they can enjoy but that would edify them and elevate their purpose in life. What I wanted to do was to get them an electronics lab kit from RadioShack. I went to the RadioShack store and there was a plethora of kits you could buy. Some were very very expensive, but they had a lot of gadgets in them. Some were medium priced. Some of them still attractively wrapped—looking very good—but were more economical. Coming from sort of a Chinese American Asian background, I succumbed to

> *my core cultural heritage and bought the more economical package. So I brought it home and the kids saw that package and were all excited and ripped it open and in five minutes they were bored to tears. Because inside of that electronics kit was wires and things like that —all the linear devices. And there was no magic. So I learned my trick. I went back to the RadioShack store and bought the Mondo kit—the exciting one—and I brought that back and they were very excited. They could build all sorts of things like a burglar alarm, and there were sirens going off—had the nonlinear devices. Those kits had the magic of electrical signals. So, for biological electrical signals, the magic is a class of molecules called ion channels. That is what we are going to be focusing on*
>
> —David T. Yue

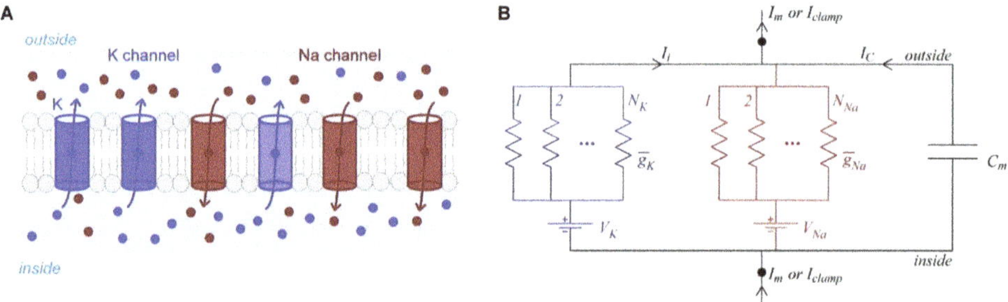

Figure 1.3. (A) Diagram showing K channels (blue) and Na channels (red) at the plasma membrane and selective transport of K^+ and Na^+ ions along the ion gradient. (B) Equivalent circuit diagram.

We have drawn it here with Na and K channels, reflecting the bias from earlier indirect experiments that Na and K transport were key to the neuronal action potential and excitability (figure 1.3). However, the formalism would be appropriate for ion channels selective for any arbitrary ion X. Generically,

$$I_X = N_X \times g_X \times \left(V_m - V_X\right) \times p_X,$$

where N_X is the number of X-type channels in a cell, g_X is the conductance of a single X-type channel when it is open, V_X is the Nernst potential for ion X, and p_X is the probability that an X-type channel is open. This equation would hold true under the assumption that individual ion channels act independently of each other.

With this generic equation for ion-channel current in mind, we can now write the overall equation that constrains the electrical behavior of the membrane of a cell:

$$I_{\text{clamp}} = I_C + I_{\text{Na}} + I_{\text{K}}$$

$$I_{\text{clamp}} = C\frac{dV_m}{dt} + N_{\text{Na}} \times g_{\text{Na}} \times (V_m - V_{Na}) \times p_{\text{Na}} \\ + N_{\text{K}} \times g_{\text{K}} \times (V_m - V_K) \times p_K. \tag{1.1}$$

1.4 Voltage-clamp tests of the HH preconception

Just before the Second World War, Alan Hodgkin learned how to manipulate current and voltage in the squid giant axon, following preparation by Kenneth S. Cole, an American (Cole and Curtis 1939). The Second World War intervened, and Alan Hodgkin became much better at electronics, working with radar. After the war, HH were the ones to exploit the squid giant axon to uncover the mysteries of ion channels. This emphasizes the advantage of theoretical experiments, even before experimental data are available.

Figure 1.4 shows the idealized setup of the voltage-clamp apparatus. Also shown are plots of the current responses to hyperpolarizing (Figure 1.4 (B) blue trace) and depolarizing (Figure 1.4 (B) red trace) step changes in V_m. In the classic 1952 papers by HH, the conventions for $\pm I$ and $\pm V$ were both reversed. In addition, the resting potential was assigned $V = 0$.

A first point to note about the records is the 'spike' of current at the instant of the voltage transition, apparent in both panels. Close inspection of these spikes validates the notion that the membrane behaves like a capacitor. The spike represents the current required to charge the membrane capacitance to the new voltage (I_C). Its duration is determined by how quickly the electronics can change V to a new constant value, because the instantaneous value of I_C is given by $C \times dV/dt$. Once V_m has been set to a new constant level, $dV_m/dt = 0$ or $I_C = 0$. Expanded timebase analysis of the spikes showed single-exponential decays with two properties that fit nicely with the idea that the membrane behaves as a capacitor (Cole and Curtis 1939). First, the area under the transients (Q) divided by the magnitude of the voltage transitions (ΔV) is invariant ($= C$). The calculated value of C gives a value of 0.9 μF cm^{-2}, a value that is about the same for most membranes studied so far. Second, the time constants of decay ($\tau = RC \sim 10\ \mu s$) are the same regardless of the magnitude of the voltage step. The measured time constant is consistent with the

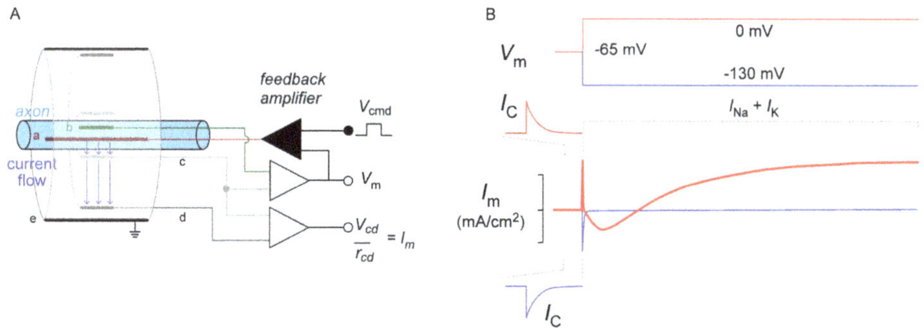

Figure 1.4. (A) Schematic of the voltage-clamp setup. The ground (e) is a cylindrical sheet. This setup ensures that the current flow is radial, and the current flow across the membrane is determined by measuring the voltage across electrodes c and d. The transmembrane voltage (V_m) is measured across electrodes b and c. A feedback amplifier and electrode deliver current to maintain V_m at the desired V_{cmd}. (B) Typical current waveforms observed in response to a change in V_m.

presence of a non-membrane resistance between the bath electrode and the current-passing electrode, a so-called 'series resistance' $R = \tau/C$ of about 7 Ω.

A second point to note is that after the spike, $I_C = 0$, so $I_{\text{clamp}} = I_{\text{Na}} + I_{\text{K}}$. This is a crucial advantage of the voltage clamp; we now have a direct measure of the current flowing through the transport mechanisms alone as a function of V_m; there is no interference from I_C.

A third point is that $I_{\text{Na}} + I_{\text{K}}$ is near zero at $V \leqslant -65$ mV but appreciable upon depolarization to 0 mV. Hence, it must be the case that the transport mechanisms are gated—i.e. they are not always on. In fact, the transport mechanisms seem to be switched on by depolarization. Hence, the term 'voltage-gated' channels.

1.4.1 Ion transport mechanisms are separable into Na and K pathways

The biphasic nature of the currents recorded (first negative, then positive) gave reason to wonder whether the early phase of current corresponded mainly to Na^+ transport and the latter phase to K^+ transport (Hodgkin and Huxley 1952d). This would fit with the Nernst potentials for these ions. Hence, there was reason to suspect that Na^+ and K^+ transport were temporally dissociated.

To bolster this idea, HH measured the current responses to a family of voltage steps as shown above. If the early phase of current corresponded to selective Na^+ transport, then a plot of the peak early current as a function of V should have passed through zero at the Nernst potential for Na^+; this is a thermodynamic mandate under the assumption of selectivity (1.2.4). Likewise, a plot of the peak late phase of current as a function of V should have passed through zero at the Nernst potential for K^+. Inspection of the early phase fitted the idea that the early current would vanish between 30 and 50 mV, i.e. near V_{Na}. The late currents increased monotonically with $V \geqslant -10$ mV, which was at least consistent with a more negative V_{K}.

Detailed plots of this kind confirm the expectations above, although there is some complication in the case of the late currents. The late current is so small near V_{K}, owing to closure of gates, that we cannot be certain that the relation passes through zero at V_{K}. The inflection in both curves is presumably due to voltage-dependent gating. So far, the data supports the idea that Na and K transports are temporally separable processes. But are the two processes functionally separable at a given instant in time (as envisioned in 1.2.4 and 1.3)? Experiments in which $[\text{Na}]_{\text{out}}$ was largely omitted supported the functional separability of the Na and K transport processes (Hodgkin and Huxley 1952c, Cole and Moore 1960). Shown in figure 1.5 is the result of such experiments. Note that the data is modified to match the modern convention, with inward current plotted as negative.

Important points are that: (1) the late phase of the current is about the same in the presence and absence of external sodium, and (2) there is a smooth rise of current in the absence of external sodium. These features are most easily reconciled with functionally separate Na and K transport systems (1.2.4 and 1.3). The sodium replacement experiment also pointed the way to obtaining separate current traces for Na and K currents. By projecting $\pm$ traces of external sodium, HH could visually subtract traces to obtain an estimate of I_{Na}. The current traces in the absence of

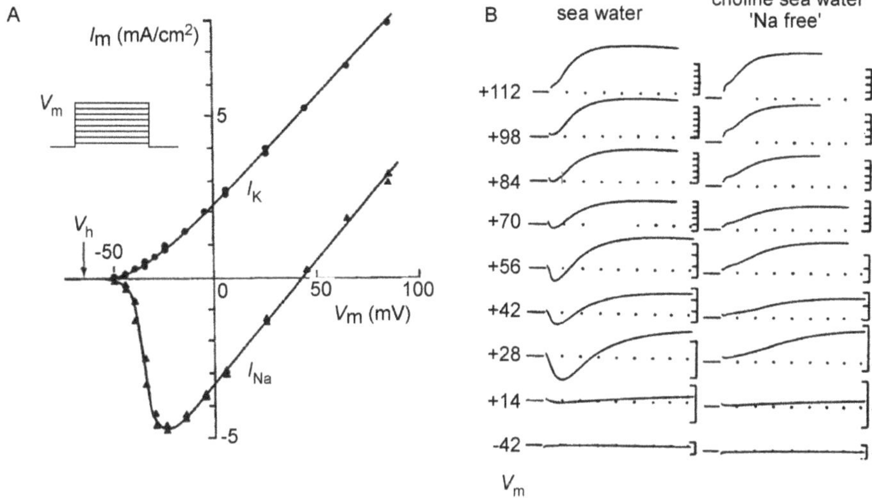

Figure 1.5. (A) Sodium and potassium currents recorded from a squid giant axon. Reproduced with permission from [Cole and Moore 1960]. Copyright, 1961, by The Rockefeller Institute. (B) Recordings from a squid giant axon with and without Na present, demonstrating the functional separation of Na^+ and K^+ currents. [Hodgkin and Huxley 1952b] John Wiley & Sons. © 1952 The Physiological Society.

external sodium could be taken as I_K. This is a slight simplification of what was actually done, because they did not omit external sodium altogether in the experiments used for quantitative analysis.

1.4.2 Conduction through open channels is ohmic and reverses at the Nernst potential

HH used a 'double-pulse' voltage protocol to show that Na and K transport mechanisms behaved ohmically with a battery offset close to the appropriate Nernst potential (Hodgkin and Huxley 1952a). Figure 1.6 shows the protocol. V_1 is kept fixed through a number of trials, while V_2 is allowed to vary. If we perform the experiment with short pulses at V_1, then most of the current is due to Na transport. In this case, the current magnitude immediately after the transition from V_1 to V_2 should be given by:

$$I_{Na} = N_{Na}g_{Na}(V_2 - V_{Na})p_{Na},$$

where p_{Na} is determined by the duration and voltage (V_1) of the first pulse. Hence, if V_1 is held fixed, p_{Na} can be considered a constant. The assumption here is that the conduction process of an open pore responds to changes in V much more quickly than the gating machinery. This seems reasonable, given the millisecond changes in current waveforms, which are indicative of gating.

Hence, if we plot I just after the transition from V_1 to V_2 (circles below) as a function of V_2, we should define a straight line with slope $N_{Na}g_{Na}p_{Na}$ and a voltage-axis intercept of V_{Na}, so long as the open pore behaves ohmically with a battery offset. The actual plot below confirms the prediction.

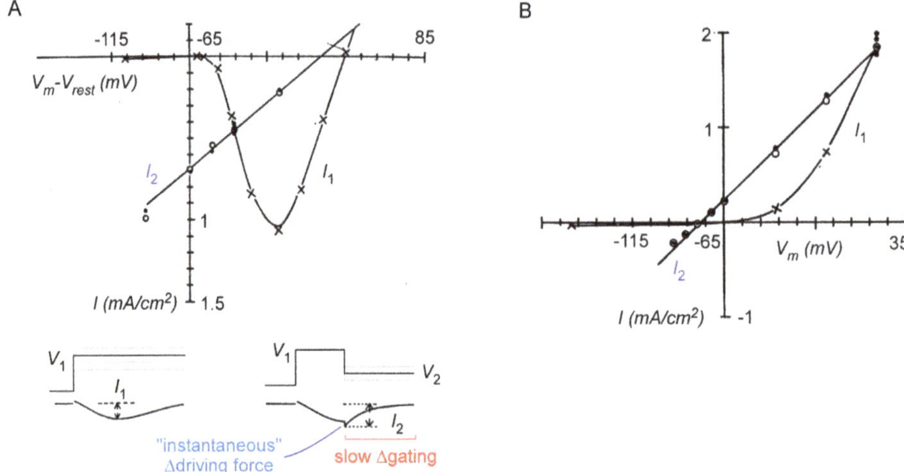

Figure 1.6. Instantaneous current–voltage relationship recorded from a squid giant axon. (A) The current (I_1) is measured from the peak response to a voltage step (V_1), producing a current–voltage relationship with a bell-shaped curve. Short-duration voltage steps (V_1) activate Na$^+$ channels without allowing inactivation. The current measured immediately after stepping to V_2 (I_1) shows a linear I–V relationship with reversal near the Na$^+$ Nernst potential. [Hodgkin and Huxley 1952a] John Wiley & Sons. © 1952 The Physiological Society. (B) Long-duration pulses isolate K$^+$ currents. The instantaneous current–voltage relationship (I_2) is again linear. [Hodgkin and Huxley 1952a] John Wiley & Sons. © 1952 The Physiological Society.

Similarly, if the experiment is repeated with long first pulses (at V_1), then the analogous plot should pertain to K transport, with a reversal near V_K. This prediction is fulfilled.

1.5 Quantitative gating mechanism

Having established the HH preconception (1.3) and developed a way to isolate I_{Na} and I_K, HH were now in a position to develop a quantitative model of gating. To do this, they transformed their isolated I_{Na} and I_K traces into waveforms of

$$G_{Na} = I_{Na}/(V - V_{Na}) = N_{Na}g_{Na}p_{Na}$$

and

$$G_K = I_K/(V - V_K) = N_K g_K p_K.$$

These are directly proportional to the open probabilities p_{Na} and p_K. These transformed waveforms are shown in figure 1.7.

1.5.1 K channel gating mechanism

To explain the K$^+$ channel waveforms, HH postulated that there were 'activation gates' called *n*-gates, all of which had to swing out of the way for the channel to be open. For simplicity, each *n*-gate was assumed to have only two possible states, a closed position (C) and a permissive position (P). Increasing depolarization could be sensed by the *n*-gates (via intrinsic charge on a voltage sensor) so as to favor

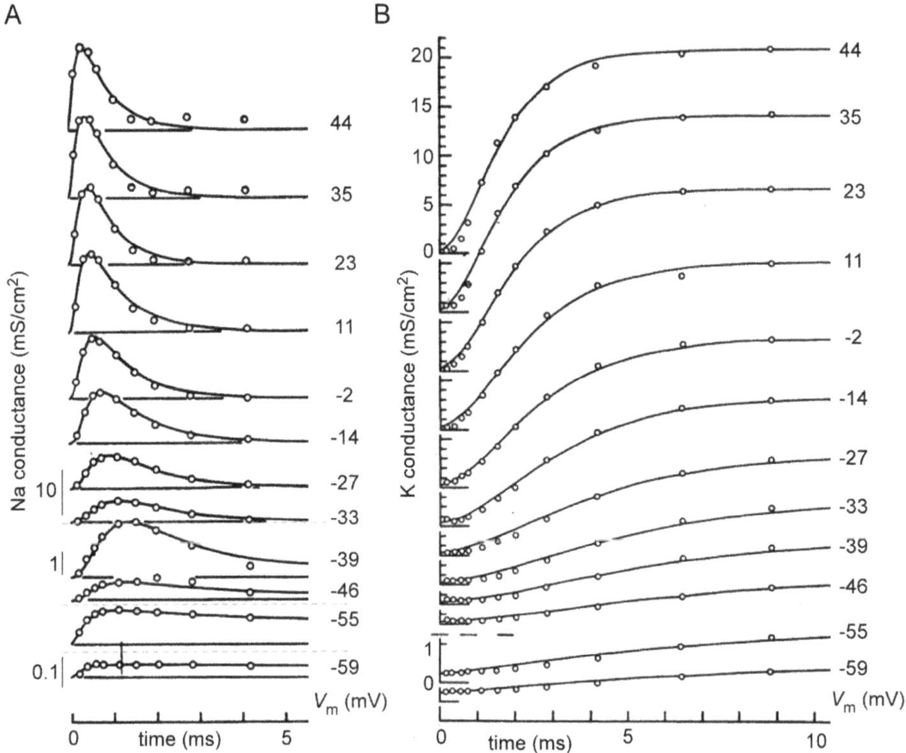

Figure 1.7. Time-dependent currents recorded from squid giant axons. (A) Sodium conductance increases rapidly following depolarization and then declines due to inactivation. This biphasic behavior supports a gating model with three activation gates (*m*) and one inactivation gate (*h*). [Hodgkin & Huxley, 1952d] John Wiley & Sons. © 1952 The Physiological Society. (B) Potassium conductance rises slowly and monotonically after depolarization, consistent with a model involving four identical activation gates (*n*) and no inactivation. [Hodgkin & Huxley, 1952d] John Wiley & Sons. © 1952 The Physiological Society. These waveforms were used to extract gating parameters and validate the Hodgkin–Huxley model of voltage-dependent conductance.

occupancy of P. Transitions between C and P were proposed to occur through first-order kinetics:

$$C_n \underset{\beta_n}{\overset{\alpha_n}{\rightleftharpoons}} P_n$$

$$\frac{dn}{dt} = \alpha_n(1 - n) - \beta_n n,$$

where n is the probability of being permissive, and α_n and β_n are believed to be instantaneous functions of V. The behavior of n in response to a voltage step from V_H to V is therefore:

$$n(t) = n(\infty, V) + [n(\infty, V_H) - n(\infty, V)]\exp\left(\frac{-t}{\tau_n(V)}\right) \tag{1.2a}$$

$$n(\infty, V) = \frac{\alpha_n(V)}{\alpha_n(V) + \beta_n(V)} \tag{1.2b}$$

$$\tau_n(V) = \frac{1}{\alpha_n(V) + \beta_n(V)}. \tag{1.2c}$$

If there were only one n-gate, we could not explain the sigmoidal delay in the activation of G_K; only a single-exponential rise is predicted by the solution above. Four n-gates turned out to give the necessary delay. Hence,

$$p_K = n^4,$$

because all four gates must be in the P position for the channel to be open (gates are assumed to be independent). Under the assumption that p_K (and therefore n) approaches one at strong depolarizations, we are justified in normalizing the G_K waveforms above by the maximum values at strong depolarization to get waveforms of p_K at various V. The initial values of such normalized plots specify $p_K(\infty, V_H) = n(\infty, V_H)^4$. Hence, $n(\infty, V_H)$ is specified. Fitting equation (1.2) to the trace at any individual voltage yields $\tau_n(V)$ and $n(\infty, V)$. These two constraints allow $\alpha_n(V) = n(\infty, V)/\tau_n(V)$ and $\beta_n(V) = (1/\tau_n(V)) - \alpha_n(V)$ to be calculated directly. These calculations lead to the following conclusions about the voltage dependence of $\alpha_n(V)$ and $\beta_n(V)$.

The data can be empirically fitted by

$$\alpha_n(V) = 0.01 \frac{(-V - 65 + 10)}{\exp\left((-V - 65 + 10)/10\right) - 1} \tag{1.3a}$$

$$\beta_n = 0.125 \exp\left(-\frac{V + 65}{80}\right) \tag{1.3b}$$

in units of m s^{-1}. Note that these equations have been corrected to the usual V convention, as opposed to the G waveforms above. Equations (1.2) and (1.3) now completely specify the gating behavior of the n-gate and thereby the K channel (figure 1.8).

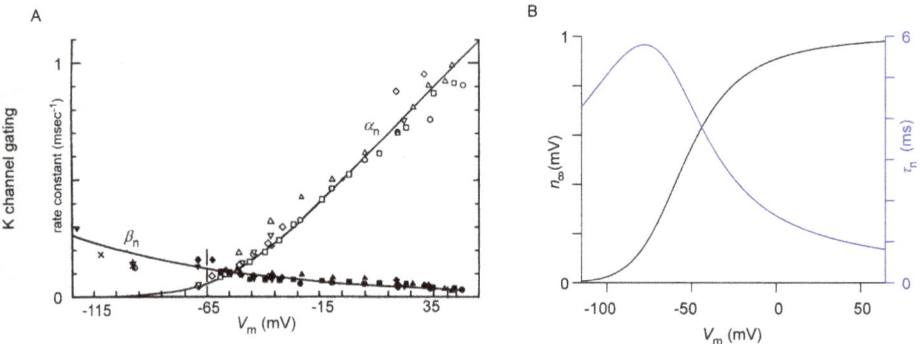

Figure 1.8. Voltage-dependent rate constants for potassium channels. (A) The rate constant α_n, increases with depolarization while β_n, decreases with depolarization. [Hodgkin & Huxley, 1952d] John Wiley & Sons. © 1952 The Physiological Society. (B) Voltage dependences of the steady-state activation variable n_∞ and the time constant τ_n.

1.5.2 Na channel gating mechanism

1.5.2.1 Estimates of τ_m, τ_h, and $m(\infty, V)$
The biphasic shape of the Na channel G_{Na} waveforms, with an initial rapid activation and subsequent slower inactivation, represents a fundamental difference in the gating properties of Na channels. The biphasic feature prompted HH to postulate that there were not only 'activation gates' called m-gates but also an 'inactivation gate' called an h-gate. All of the gates had to swing out of the way for the channel to be open. For simplicity, each h- and n-gate was assumed to have only two possible states: a closed position (C) and a permissive position (P). Increasing depolarization could be sensed by the m-gates (via intrinsic charge on a voltage sensor) to favor occupancy of P. To account for inactivation, increasing depolarization could be sensed by the h-gates to favor occupancy of C. Transitions between C and P were proposed to occur through first-order kinetics:

$$C_m \underset{\beta_m}{\overset{\alpha_m}{\rightleftharpoons}} P_m$$

$$\frac{dm}{dt} = \alpha_m(1 - m) - \beta_m m,$$

where m is the probability of being permissive for the m-gate. The rate constants α and β are believed to be instantaneous functions of V. A similar relation holds true for the h-gate:

$$C_h \underset{\beta_h}{\overset{\alpha_h}{\rightleftharpoons}} P_h$$

$$\frac{dh}{dt} = \alpha_h(1 - h) - \beta_h h,$$

where h is the probability of being permissive for the h-gate.

The behaviors of m and h in response to a voltage step from V_H to V are therefore:

$$m(t) = m(\infty, V) + [m(\infty, V_H) - m(\infty, V)]\exp\left(\frac{-t}{\tau_m(V)}\right) \tag{1.4a}$$

$$m(\infty, V) = \frac{\alpha_m(V)}{\alpha_m(V) + \beta_m(V)} \tag{1.4b}$$

$$\tau_m(V) = \frac{1}{\alpha_m(V) + \beta_m(V)} \tag{1.4c}$$

$$h(t) = h(\infty, V) + [h(\infty, V_H) - h(\infty, V)]\exp\left(\frac{-t}{\tau_h(V)}\right) \tag{1.4d}$$

$$h(\infty, V) = \frac{\alpha_h(V)}{\alpha_h(V) + \beta_h(V)} \tag{1.4e}$$

$$\tau_h(V) = \frac{1}{\alpha_h(V) + \beta_h(V)}. \tag{1.4f}$$

Three m-gates were required to fit the initial activation kinetics, but a single h-gate sufficed to fit the declining inactivation phase of current. Because all gates must be in P for the Na channel to be open, we have that

$$p_{Na} = m^3 h.$$

To constrain α and β, it was assumed that $m(\infty)$ approached one at strong depolarizations and approached zero at strong hyperpolarizations ($V = V_H$; this fits with the lack of I_{Na} at hyperpolarized V_H). Likewise, it was assumed that $h(\infty)$ approached one at strong hyperpolarizations and approached zero at strong depolarizations ($V \geqslant -35\,\mathrm{mV}$; this fits with the lack of I_{Na} after maintained depolarization). Under these assumptions,

$$G_{Na} = \underbrace{N_{Na}g_{Na}m(\infty, V)h(\infty, V_H)}_{\overline{G_{Na}}}\left(1 - \exp\left(-\frac{t}{\tau_m}\right)\right)^3 \exp\left(-\frac{t}{\tau_h}\right),$$

where $h(\infty, V_H) \sim 0.5$ was determined by a prepulse-inactivation protocol described in the next section. Under the assumption that $m(\infty, V) \sim 1$ at strong depolarization, fits to traces elicited by very positive voltages yielded $(N_{Na}g_{Na})$, τ_h, and τ_m. With knowledge of $(N_{Na}g_{Na})$, fits to traces elicited by less positive voltages yielded estimates of $m(\infty, V)$, τ_h, and τ_m.

Given the specification of $m(\infty, V)$ and τ_m as a function of voltage, HH could directly calculate $\alpha_m(V) = m(\infty, V)/\tau_m(V)$ and $\beta_m(V) = (1/\tau_m(V)) - \alpha_m(V)$.

These calculations lead to the following conclusions about the voltage dependence of $\alpha_m(V)$ and $\beta_m(V)$. The data can be empirically fit by:

$$\alpha_m(V) = 0.1\frac{(-V - 65 + 25)}{\exp\left((-V - 65 + 25)/10\right) - 1} \tag{1.5a}$$

$$\beta_m = 4\exp\left(-\frac{V + 65}{18}\right) \tag{1.5b}$$

in units of ms^{-1}. Note that these equations have been corrected to the usual V convention (figure 1.9).

So far, we only have determinations of τ_h at various V. To obtain estimates of $h(\infty, V)$, which are required to specify α_h and β_h independently, we use a prepulse-inactivation protocol, shown below. The peak I_{Na} is given by

$$I_{Na} = N_{Na}g_{Na}(-15 - V_{Na})m(t_{max})^3 \exp\left(-\frac{t_{max}}{\tau_h(-15)}\right)h(\infty, E_{pre}).$$

I_{Na} obtained with very hyperpolarized prepulses yields

$$I_{Na,\,max} = N_{Na}g_{Na}(-15 - V_{Na})m(t_{max})^3 \exp\left(-\frac{t_{max}}{\tau_h(-15)}\right),$$

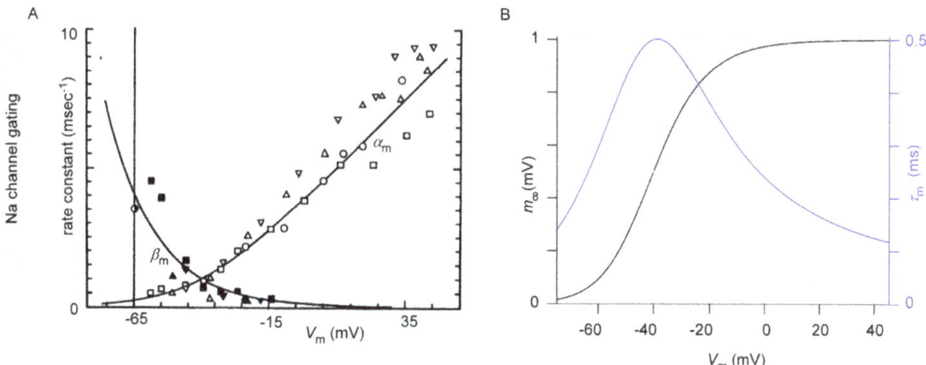

Figure 1.9. Voltage-dependent rate constants for sodium channel opening. (A) The rate constant α_m increases with depolarization, while β_m decreases with depolarization. [Hodgkin & Huxley, 1952d] John Wiley & Sons. © 1952 The Physiological Society. (B) Voltage dependences of the steady-state activation variable n_∞ and the time constant τ_n.

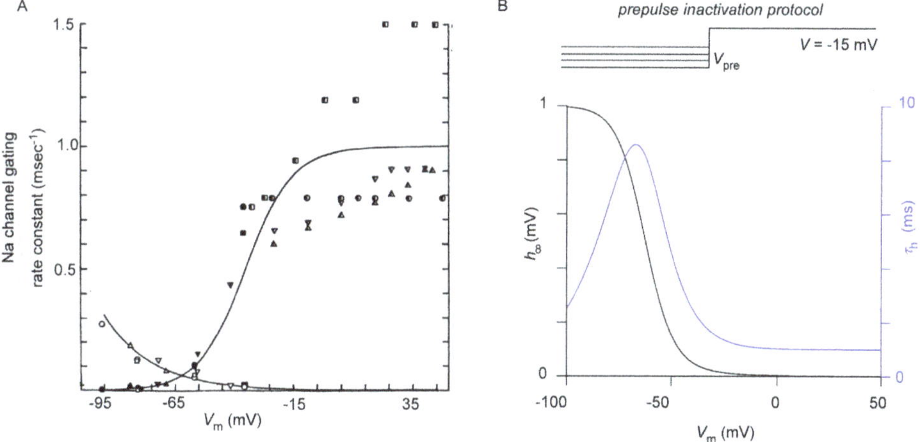

Figure 1.10. Voltage-dependent rate constants for sodium channel inactivation. (A) The rate constant α_h decreases with depolarization, while β_h increases with depolarization. [Hodgkin & Huxley, 1952d] John Wiley & Sons. © 1952 The Physiological Society. (B) Voltage dependences of the steady-state activation variable h_∞ and the time constant τ_h.

where $h(\infty, E_{\mathrm{pre}})$ is assumed to be unity. If we normalize I_{Na} by $I_{\mathrm{Na,\ max}}$, we therefore obtain a direct estimate of $h(\infty, V) = I_{\mathrm{Na}}/I_{\mathrm{Na,\ max}}$.

We have now specified τ_h and $h(\infty, V)$ at various V. With this information, HH could directly calculate $\alpha_h(V) = h(\infty, V)/\tau_h(V)$ and $\beta_h(V) = (1/\tau_h(V)) - \alpha_h(V)$ (figure 1.10).

These calculations lead to the following conclusions about the voltage dependences of $\alpha_h(V)$ and $\beta_h(V)$. The data can be empirically fitted by

$$\beta_h(V) = \frac{1}{\exp\left((-V - 65 + 30)/10\right) + 1} \tag{1.6a}$$

$$\alpha_h = 0.07 \exp\left(-\frac{V + 65}{20}\right) \tag{1.6b}$$

in units of m s^{-1}. Note that these equations have been corrected to the usual V convention.

Equations (1.4)–(1.6) thereby enable the complete specification of Na channels.

1.5.3 'Proof' of the HH gating mechanism

The gating equations (1.2)–(1.6), when plugged into the complete cell model (equation (1.1)), did a beautiful job of predicting action-potential phenomena.

1.6 Enduring features of the HH viewpoint

Although the specifics of the HH view have been shown to be incorrect, certain general features have stood the test of time and form the basis of the modern understanding of ion channels, as follows: (1) Separate types of channels specific to one type of ion or another; (2) separation of gating and permeation; (3) ohmic + battery conduction is not a bad description; (4) separation of voltage activation and voltage inactivation; (5) multiplicity of molecular charge sensors (S4 motifs) and thereby activation gates; (6) only one inactivating particle required.

The HH equations are also still useful as a simple quantitative formulation that does a reasonable job of predicting channel behavior in cellular models. The ease with which the core equations can be calculated makes this formalism a favorite for models of cells or networks of cells.

1.7 A qualitative approach to the reconstruction of the action potential

The Noble approximation is a brilliant example of graphical analysis (Noble 1962). The fully quantitative equations describing action-potential genesis are somewhat daunting:

$$CdV/dt = -I_{\text{Na}} - I_{\text{K}}$$
$$dV/dt|_{t=\sigma} = [-1/C] \cdot [N_{\text{Na}} \cdot m(V(t))^4 \cdot h(V(t)) \cdot g_{\text{Na}} \cdot (V(\sigma) - V_{\text{Na}})$$
$$+ N_{\text{K}} \cdot n(V(t))^4 \cdot g_{\text{K}} \cdot (V(\sigma) - V_{\text{K}})],$$

where $V(t)$ represents the entire history of V up until the current instant of time σ, and m, n, and h need to be determined by solving their differential equations for that entire history. Yikes! The magic behind Noble's quasi-quantitative approximation is this: m equilibrates with changes in voltage so much faster than h or n that we can consider an approximation where: (1) only m is a function of time, (2) m takes on its steady-state value with respect to V at the current instant of time $\sigma(V(\sigma))$. h, and (3) n can be considered constant with respect to time. Hence,

$$dV/dt|_{t=\sigma} \sim [-1/C] \times [N_{\text{Na}} \times m(\infty, V(t))^4 \times h \times g_{\text{Na}} \times (V(\sigma) - V_{\text{Na}})$$
$$+ N_{\text{K}} \times n^4 \times g_{\text{K}} \times (V(\sigma) - V_{\text{K}})].$$

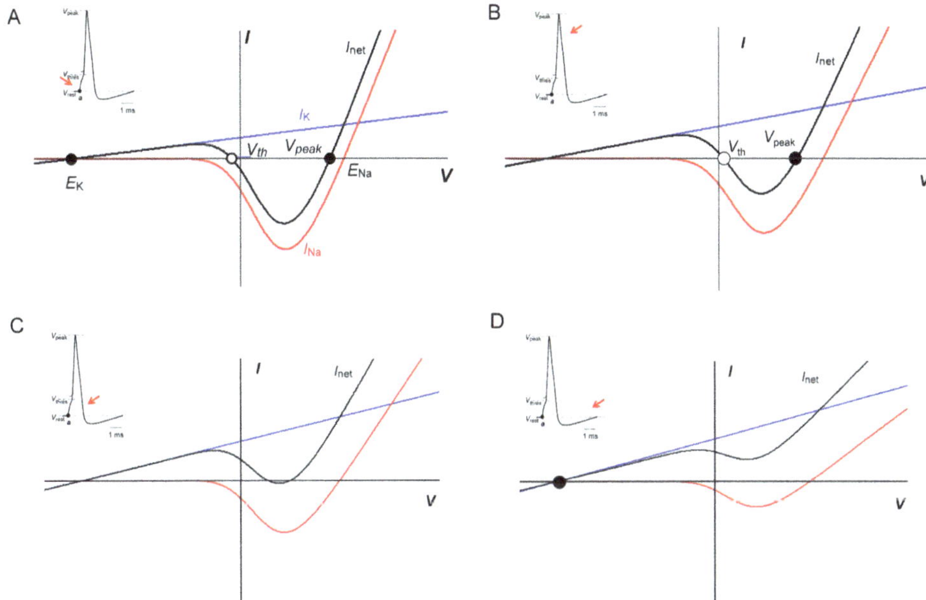

Figure 1.11. Noble approximation interpretation of action potentials. (A) Graphical construction of the quasi-instantaneous current–voltage relation (dashed), composed of the instantaneous current–voltage relation for K channels, the Na channel curve obtained with inactivation held constant, and m allowed to adopt $m(\infty,V)$ values. Zero crossings of the dashed curve have special meanings, as described in the text. (B–D) Family of quasi-instantaneous current–voltage relations as the action potential progresses. As the action potential ensues, the n-gate activates more and the h-gate inactivates more, but three stable points remain; eventually, they collide, leaving only a single stable point, the resting potential. This is the refractory period.

Thus, $dV/dt|_{t=\sigma}$ is approximately proportional to the black curve in figure 1.11, which is called the quasi-instantaneous current–voltage relation for the cell. The zero-crossing points of the dashed relation are of special significance because they are voltages where the cell may tend to reside if the points are 'stable.' Stability is determined by the slope of the crossing as follows:

- A positive slope through the $I = 0$ crossing implies a stable point. Spontaneous depolarization produces outward current, which repolarizes. Spontaneous hyperpolarization produces inward current, which depolarizes.
- A negative slope through the $I = 0$ crossing implies an unstable point. Spontaneous depolarization produces inward current, which begets more depolarization. Spontaneous hyperpolarization produces outward current, which begets more hyperpolarization.

With this, we can interpret the threshold potential, the peak of the action potential, and the refractory period (figure 1.11).

As Na channel inactivation takes place, the quasi-instantaneous current–voltage relation tracks through a family of curves, as shown in figures 1.11(B)–(D), driven by the inactivation of Na channels and the ongoing activation of K channels. Loss of the V_{peak} stable point in figure 1.11(C) signals the advent of the relation in figure 1.11(D).

From this instant on, the only stable point is that for the resting potential, and the cell inexorably repolarizes towards that target. This graphical approach is powerful, rich in intuition, and a useful adjunct to the full-bore computer simulation that we will unwrap in the next chapter. Graphical analysis can sometimes explain *why* something strange and wonderful is happening, whereas brute-force computer simulation can only demonstrate that something strange and wonderful will happen. If the equations being simulated are sufficiently complex, then the peculiar behavior of the computer simulation may be every bit as mystical as the real-life phenomena being simulated.

References

Cole K S and Curtis H J 1939 Electric impedance of the squid giant axon during activity *J. Gen. Physiol.* **22** 649–70

Cole K S and Moore J W 1960 Ionic current measurements in the squid giant axon membrane *J. Gen. Physiol.* **44** 123–67

Goldman D E 1943 Potential, impedance, and rectification in membranes *J. Gen. Physiol.* **27** 37–60

Hille B 2001 *Ion Channels of Excitable Membranes* (Sunderland, MA: Sinauer)

Hodgkin A L and Huxley A F 1952a The components of membrane conductance in the giant axon of Loligo *J. Physiol.* **116** 473–96

Hodgkin A L and Huxley A F 1952b Currents carried by sodium and potassium ions through the membrane of the giant axon of Loligo *J. Physiol.* **116** 449–72

Hodgkin A L and Huxley A F 1952c The dual effect of membrane potential on sodium conductance in the giant axon of Loligo *J. Physiol.* **116** 497–506

Hodgkin A L and Huxley A F 1952d A quantitative description of membrane current and its application to conduction and excitation in nerve *J. Physiol.* **117** 500–44

Hodgkin A L, Huxley A F and Katz B 1952 Measurement of current–voltage relations in the membrane of the giant axon of Loligo *J. Physiol.* **116** 424–48

Hodgkin A L and Katz B 1949 The effect of sodium ions on the electrical activity of giant axon of the squid *J. Physiol.* **108** 37–77

Noble D 1962 A modification of the Hodgkin–Huxley equations applicable to Purkinje fibre action and pace-maker potentials *J. Physiol.* **160** 317–52

IOP Publishing

Ion Channel Gating and Mechanisms

Chapter 2

The big deal about single-channel information

2.1 Introduction

What can single-channel records tell us about gating? The purpose of this chapter is to illustrate how single-channel records can distinguish between very different gating mechanisms that nevertheless demonstrate similar behavior at the level of macroscopic currents (the level studied by Hodgkin and Huxley (HH)). To exploit the new gating information contained in single-channel records, we need to think about channels in a new way: as a special class of stochastic process, a Markov process. We will only develop a bare-bones primer on Markov processes, just enough to understand a case example taken from the annals of Na channel gating: Aldrich RW, Corey DP, and Stevens CF (1983), 'A reinterpretation of mammalian sodium channel gating based on single-channel recording.' *Nature* **306** 436–41 (Aldrich *et al* 1983). This initial foray into single-channel analysis will motivate a more complete treatment of Markov process analysis of ion channel gating in subsequent chapters.

2.2 HH gating viewed in the framework of multistate Markov models

The HH viewpoint is beautiful; why bother to change? Here, we recast the HH gating equations into a different framework, that of multistate Markov models. We do so for two reasons. First, the HH model of gating is only a special case of more general gating mechanisms, and these general mechanisms can only be represented by frameworks such as Markov models. Second, the core equations for Markov models turn out to be entirely analogous to those for linear, time-invariant systems, those so familiar to electrical and chemical engineers. Hence, the enormous wealth of quantitative insight that has been developed for linear systems can be appropriated by the Markov view of channel gating.

doi:10.1088/978-0-7503-2388-8ch2   2-1

2.2.1 Two-position gates imply a 'state' view of protein conformational changes

The assumption that a channel gate can only assume two positions, closed (C) and permissive (P), implies a state view of protein conformational changes. This need not be true. The state view is a simplification of a possible, more generalized case. The difference between the 'general' and 'state' viewpoints is illustrated by the diagram in figure 2.1.

The top row of graphs shows the free energy of the channel gate plotted as a function of its conformation, here represented simplistically as a linear reaction coordinate (in actuality, it would be a multidimensional vector). In the general view, there are several distinguishing features: (1) the free-energy profile is smooth, without sharply defined peaks and valleys; (2) based on intuition from the Boltzmann distribution (which only applies to the channel at equilibrium), we would anticipate that the probability of observing the channel in conformation X would be appreciable for many values of X. The anticipated equilibrium probabilities of the channel residing in various conformations are shown in the plot below. (3) The equations of motion for transitions among the various conformations are complex and can have diffusion-like terms. These are beyond the scope of this chapter but are nevertheless required for the most general conception of protein conformational change.

By contrast, in the state approximation shown in the right column, the situation is considerably simplified: (1) the free-energy profile is distinguished by two deep and sharply delimited valleys separated by a tall peak; (2) hence, the probability of observing the channel in various conformations is concentrated in two loci, corresponding to states C and P; (3) the transitions between C and P can now be described by the simple chemical reaction scheme

$$C \underset{\beta}{\overset{\alpha}{\rightleftharpoons}} P$$

or

$$\frac{dC}{dt} = -\alpha C + \beta P$$

$$\frac{dP}{dt} = +\alpha C - \beta P,$$

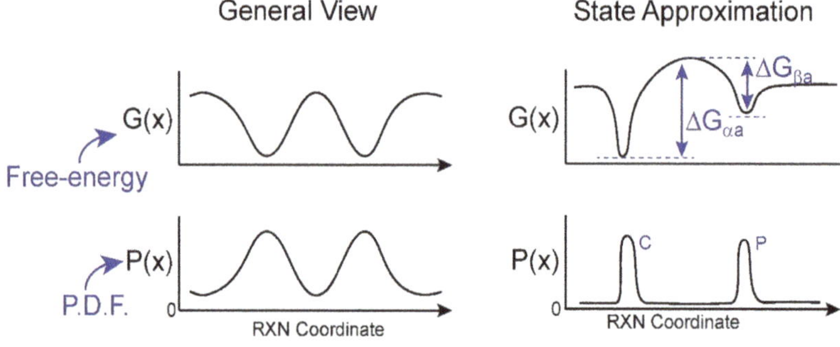

Figure 2.1. Graphs depicting the general and state approximation schemes of channel gating. The general view shows a smooth free-energy landscape with multiple conformations, while the state view depicts two sharply defined states (C and P) separated by a high energy barrier. P.D.F. denotes probability density function.

where C and P above stand for the probability of observing the gate in state C or state P, respectively. The new information from the state view comes with the relation between the free-energy profile and the rate constants α and β. According to Eyring rate theory (Eyring 1935) (which is only strictly true for gases),

$$\alpha = \kappa \frac{kT}{h} \exp\left(-\frac{\Delta G_{\alpha a}}{RT}\right) \tag{2.1a}$$

$$\beta = \kappa \frac{kT}{h} \exp\left(-\frac{\Delta G_{\beta a}}{RT}\right), \tag{2.1b}$$

where κ is the transmission coefficient (generally assumed to be unity), kT/h is $6 \times 10^{12}\,\mathrm{s}^{-1}$ at 20 °C, and the ΔG terms are activation energies. The conditions required for the Eyring approximation to hold include sharply defined valleys (well-defined states) and sufficiently large activation barriers (a few RT). While the conditions required for strict application of the Eyring rate theory may not always be met for channel proteins, these expressions (especially the exponential dependence on activation energies) have proven to be useful approximations in understanding channel function.

2.2.2 K-channel gating: from four gates to five channel states

Because the gates can only assume two states, the channel gating mechanism can only adopt a finite number of states, defined by the positions of the various gates. In the case of the K channel, with four n-gates, the state of the K channel can be specified by a 4×1 vector, each of whose elements can be zero or one, depending upon whether a given gate is in the C or P state, respectively. Because transitions between the two states of a gate are specified by rate constants, so too must transitions between channel states. Overall, channel gating can be recast as the five-state chemical reaction model shown in panel **B** of the following figure, as outlined below.

First, since there are four gates with two states, the channel as a whole must have 2^4 states, as diagrammed in panel A. From equation (2.1), the overall equations describing interstate transitions could be written as:

$$\frac{d(C_{1A})}{dt} = -4\alpha(C_{1A}) + \beta(C_{2A} + \ldots + C_{2D})$$

$$\frac{d(C_{2A} + \ldots + C_{2D})}{dt} =$$
$$4\alpha C_{1A} - (\beta + 3\alpha)(C_{2A} + \ldots + C_{2D}) + 2\beta(C_{3A} + \ldots + C_{3F})$$

$$\frac{d(C_{3A} + \ldots + C_{3F})}{dt} =$$
$$3\alpha(C_{2A} + \ldots + C_{2D}) - (2\beta + 2\alpha)(C_{3A} + \ldots + C_{3F}) + 3\beta(C_{4A} + \ldots + C_{4D})$$

$$\frac{d(C_{4A} + \ldots + C_{4D})}{dt} =$$
$$2\alpha(C_{3A} + \ldots + C_{3F}) - (3\beta + \alpha)(C_{4A} + \ldots + C_{4D}) + 4\beta O_A$$

$$\frac{d(O_A)}{dt} = \alpha(C_{4A} + \ldots + C_{4D}) - 4\beta(O_A),$$

where the names of states and the probabilities of occupying them are used synonymously. If the n-gates are identical and independent, then all C_{2X}, C_{3X}, and C_{4X} states are functionally the same. We might as well define a new 'compound' state C_2 that refers to the channel when it occupies any state C_{2A}, ... , C_{2D}. We can define compound states C_3 and C_4 in a similar manner. We might as well rename state C_{1A} as C_1, and O_4 as O, because there is only one state in each of these categories. With these redefinitions, the differential equations for the K channel can be simplified to:

$$\frac{d(C_1)}{dt} = -4\alpha(C_1) + \beta(C_2) \tag{2.2a}$$

$$\frac{d(C_2)}{dt} = +4\alpha(C_1) - (\beta + 3\alpha)(C_2) + 2\beta(C_3) \tag{2.2b}$$

$$\frac{d(C_3)}{dt} = +3\alpha(C_2) - (2\beta + 2\alpha)(C_3) + 3\beta(C_4) \tag{2.2c}$$

$$\frac{d(C_4)}{dt} = +2\alpha(C_3) - (3\beta + \alpha)(C_4) + \frac{C}{\beta}(O) \tag{2.2d}$$

$$\frac{d(O)}{dt} = \alpha(C_4) - 4\beta(O). \tag{2.2e}$$

This set of equations is equivalent to the five-state chemical reaction scheme in panel B (figure 2.2).

A similar line of argument shows that the HH Na channel gating model can be viewed as the eight-state Markov model shown in figure 2.3 (Hille 2001). As mentioned before, the power of the Markov model formulation is twofold. First, the Markov representation can accommodate a general class of gating mechanisms that cannot be represented by a system of independent gates. If we broke the strict symmetry regarding the relations between rate constants in both the K and Na channels, the gating mechanism would be 'valid,' but there would be no analogous representation as a system of independent gates. If, for example, the voltage sensors of a channel do not operate independently, the Markov formulation could account for 'cooperativity.' Second, the equations for the Markov representation (illustrated above) are identical to those of a linear, time-invariant system. Hence, all the insights of linear system theory can be immediately appropriated for our purposes.

2.3 Single-channel records: a first look

2.3.1 Channels manifest all-or-none conductance

Shown below in figure 2.4 are records from a single cardiac Na channel. (see Box 2.1) It is clear that the channel can only demonstrate two levels of current at a single V: zero or i (the so-called unitary current amplitude). Gating concerns the conformational changes underlying switching between the conducting and nonconducting

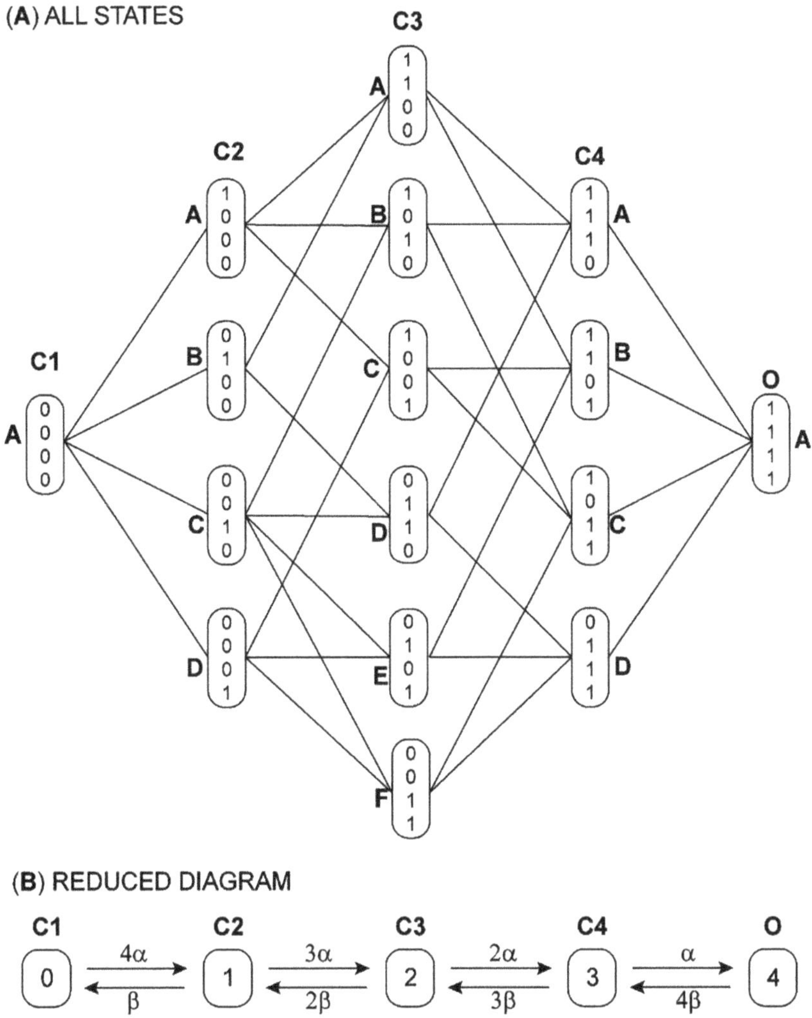

Figure 2.2. Kinetic states of the HH K$^+$ channel (Hodgkin and Huxley 1952). (A) If the K channel had four n-gates that could be open (one) or shut (zero), there would be 16 substates of the system with 32 permitted transitions. (B) If the n-gates were identical, many substates would become equivalent, and the system could be described by five major states with the transition rate constants shown. The state numbers correspond to the numbers of open gates. Only state four actually conducts ions. Reproduced with permission from [Hille, 2001].

levels. Permeation concerns the processes specifying the size of i. The records below gate in the manner predicted by the HH model. Channels first open quickly, as expected from the rapid kinetics of m-gates. Once open, the channel stays open, as expected from the high peak open probability ($m \rightarrow 1$ before h starts declining) predicted by the model. Finally, once the channel shuts, it does not reopen, as predicted from the long-lived closure of the h-gate. Not all Na channels gate like this, as illustrated by Aldrich *et al* (1983) (figure 2.4).

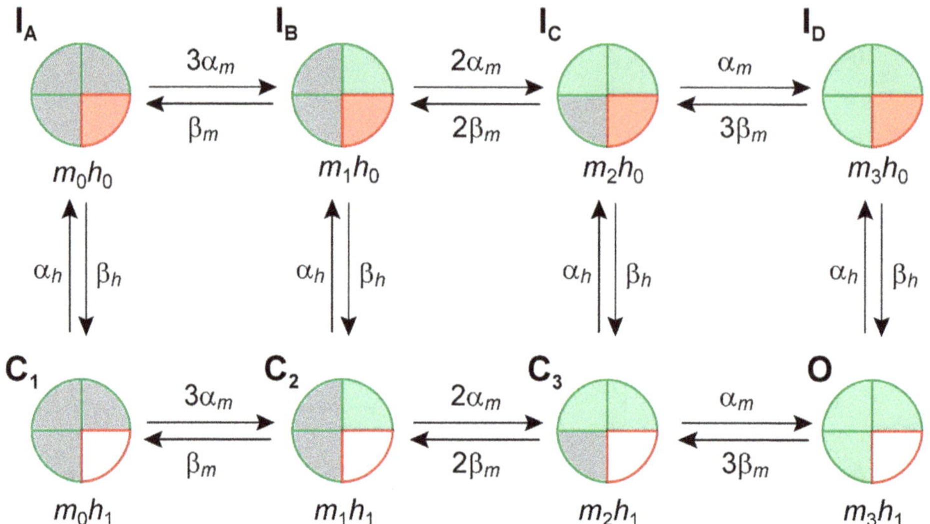

Figure 2.3. Kinetic states of the HH Na$^+$ channel (Hodgkin and Huxley 1952). Eight major gating states of the HH Na$^+$ channel model, derived from combinations of m and h-gates. Only the state with all m-gates (green outline) open (closed: gray fill; open: green fill) and the h-gate (red outline) permissive (non-permissive: red fill; permissive: white fill) conducts ions. Based on Hille (2001).

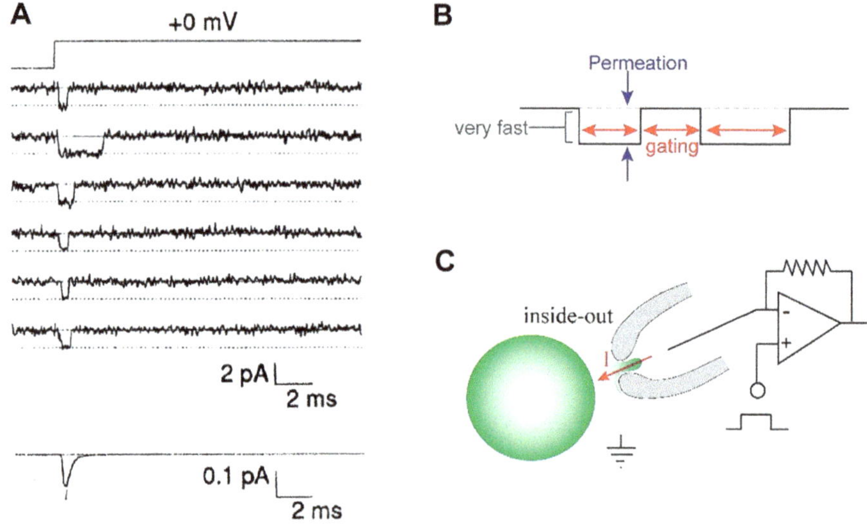

Figure 2.4. Single-channel Na$^+$ recordings. (A) Single-channel recordings from cardiac Na$^+$ channels showing all-or-none conductance. From [Yue *et al* 1989]. Reprinted with permission from AAAS. (B) Channel gating scheme. Channels open rapidly, remain open briefly, and then inactivate without reopening, consistent with the predictions of HH. (C) Inside-out patch-clamp configuration.

Box 2.1. Bruce Lee and the broken beer bottles—analogy for the patch clamp.

This story here encapsulates the magic that turns that jagged [pipe] into something smooth. So, back to 'Enter the Dragon' with Bruce Lee. And I think this is the number one bad guy or the chief evil doer. He is really bad. And bad guys although they fight really well they fight dirty. So, this guy is kind of fighting dirty. This guy earlier in the movie, actually killed Bruce Lee's sister in cold blood. He knows that so they are coming out for the match and so it is not just competition. It is personal. And the way these matches are they both hold their hands up there and no one tells them when the match starts—it is whoever moves first. That is when it starts. You have to have fast reflexes. And this bad guy is really confident because before the scene here, just to show off he had this like huge board and he goes in there and 'hoof!' and then just splinters it into bunch of little pieces just to intimidate Bruce Lee. And he just looks at that and says 'hmm, that is interesting.' And so they are up there like that. And Bruce lee who is super-fast, wow, he moves so fast. This guy couldn't react he was completely struck to the ground. They do that three times. So this bad guy is not going to lose fair and square. He is mad. So what he does—he goes to the side and he gets some bottles. They were like beer bottles. He breaks them and then Bruce Lee is walking backwards because he thinks he won and this guy is going to come and attack him from the back. He is going to fight him unfairly. I won't tell you what happens. Imagine that broken beer bottle there. So how will you make it smooth and make this a peaceful outcome. This is what Erwin Neher and Bert Sakmann did. They subjected this to high heat wire. It is like you have the broken beer bottle and stick it in a Bunsen burner. And it melts and gets gooey. Then when you take it away, it is smooth. So by fire polishing their pipets they made it very smooth. And then on a nerdy weekend when they were working away, he kissed up to a cell with these polished electrodes and formed the first gigaseal. For that they got the Nobel Prize.

—David Yue

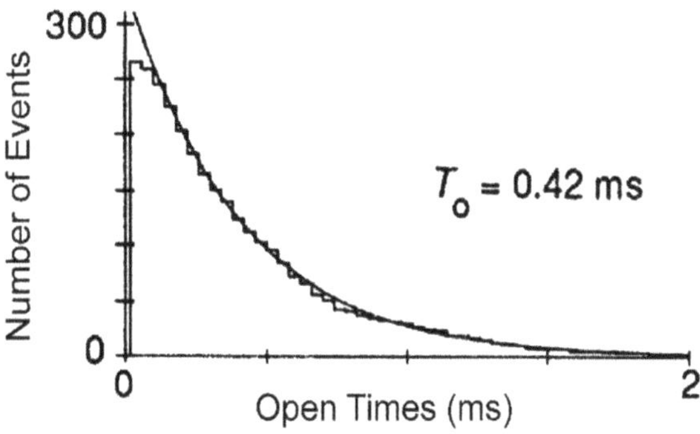

Figure 2.5. Histogram of open durations for cardiac Na^+ channels showing a single-exponential decay, indicating memoryless behavior and supporting a Markov process model for gating. From [Yue *et al* 1989]. Reprinted with permission from AAAS.

2.3.2 The open state of many channels suggests that gating is a Markov process

If we tally all the open events that last $>t$ ms and plot this as a function of t, we obtain the histogram in figure 2.5 below, which is derived from cardiac Na channels. The histogram describes a single-exponential function, shown by the smooth curve.

If we renormalize the plot so that the value of the exponential at zero time becomes unity, the resulting plot is equivalent to $P_{L_o}(t)$, the probability that the duration of an opening will be longer than t (figure 2.5). The important feature of the histogram is its single-exponential shape. This feature is observed for many ionic channels, not only the Na channel. The significance of this feature is that single-exponential lifetime distributions for a single state (O, in this case) are synonymous with exit processes that are memoryless. By memoryless, we mean that the subsequent behavior of the system is completely specified by the present state of the system. By definition, such a system behaves according to a Markov process.

It turns out that the analogous histogram for the lifetime of nonconducting periods is seldom described by a single-exponential function. Rather than indicating that exit processes from closed states are not memoryless (and therefore non-Markovian), it seems more likely that there are multiple, interconnected closed states (figure 2.5).

2.4 The power of single-channel information

The HH equations do a fairly good job of predicting the macroscopic currents flowing through voltage-gated Na and K channels. They are sufficient to explain propagating action-potential behavior. Are the HH equations then correct? Do they correspond to molecular reality? Figure 2.6 confronts us with a sobering challenge (Aldrich 1986). Shown in the figure are three different molecular mechanisms of gating, all of which produce the same shaped waveforms of macroscopic current. In other words, HH analysis would give the same sort of parameters for all three mechanisms, but only the mechanism in panel (a) would be accurately reflected by the equations.

The saving grace is this: if we were able to observe the single-channel records arising from the different mechanisms of gating, the visual impression is that we would be able to make the important distinctions. The Aldrich *et al* paper (Aldrich *et al* 1983) elaborates a beautiful case example in which Na channel gating (in some neuronal cells, but not all) can be rigorously shown to gate according to the mechanism in panel (c). This mechanism is, in some sense, the exact opposite of the classic HH model (panel (a)).

The differences in the gating mechanisms are of more than academic interest. First, to understand the molecular operation of the channel at the level of structure–function relationships, it is necessary to have a proper molecular mechanism for gating, not just a macroscopic mimic. Second, there can be enormous differences in physiological function. In the case of calcium channels, the different gating mechanisms would produce very different calcium concentration profiles in local domains near the channels, leading to very different neurotransmitter release profiles at presynaptic terminals.

Before proceeding, it is worth presenting a few mathematical and probabilistic preliminaries.

2.5 Barebones continuous-time Markov processes

A finite-state, continuous-time Markov process pertains to a system that can adopt one of a finite number of states at time t (a real number, as opposed to an integer), and the subsequent probabilities of occupying various states are completely specified by the state

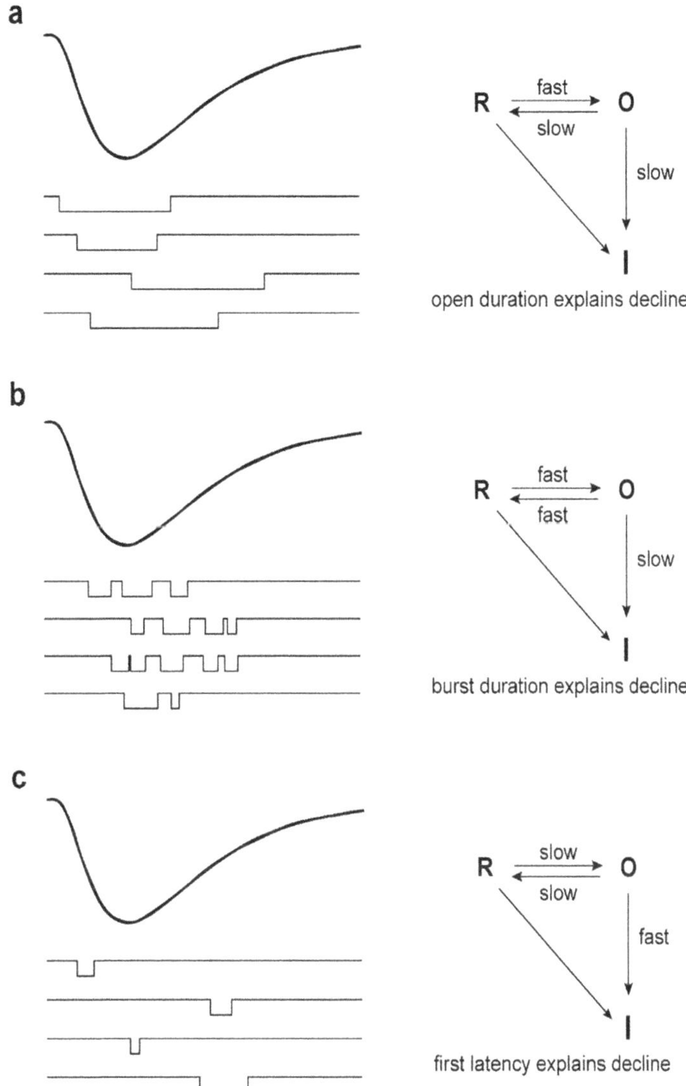

Figure 2.6. Three possibilities for sodium-channel gating that predict identical macroscopic sodium currents but different single-channel behaviors. Reprinted from [Aldrich, 1986], with permission. Copyright © 1986. Published by Elsevier Ltd.

occupied at time t (the so-called 'Markovian' assumption). Hence, the process is memoryless. It can be shown that the assumption of memoryless behavior is synonymous with exponentially distributed lifetimes in each state. This feature allows us to establish a basic 'grammar' of operational features for transitions among individual states.

2.5.1 Probability basics: random variables, probability density functions, probability distribution functions, expectation values, and binomial distribution

A random variable $X(t)$ is a number whose value is specified by a deterministic set of rules based on the state occupied by a system at time t. Transitions among states occur in a stochastic fashion. For example, the current flowing through a single channel as a function of time $i(t)$ is a random variable based on the state occupancy of a channel. When the channel is in state O, we assign a value of i to $i(t)$. When any other state is occupied, we assign a zero value to $i(t)$.

The probability density function for a random variable $X(t)$ is called $p_{X(t)}(\sigma)$ (figure 2.7). The probability that $X(t)$ is between A and $A + \Delta X$ is given by

$$\int_{\sigma=A}^{A+\Delta X} p_{X(t)}(\sigma)d\sigma.$$

In other words, for small $\Delta\sigma$, $p_{X(t)}(\sigma)\Delta\sigma$ approximately equals the probability that $X(t)$ is between σ and $\sigma + \Delta\sigma$.

A related entity is the distribution function of $X(t)$, defined as $\overline{P_{X(t)}}(\sigma) \equiv \int_{\gamma=-\infty}^{\sigma} p_{X(t)}(\gamma)d\gamma$ (figure 2.7). Another way of viewing the interrelation is $d\,\overline{P_{X(t)}}(\sigma)/d\sigma = p_{X(t)}(\sigma)$. The complement of the distribution function of $X(t)$ is $P_{X(t)}(\sigma) = 1 - \overline{P_{X(t)}}(\sigma)$ (figure 2.7). Hence, $-dP_{X(t)}(\sigma)/d\sigma = p_{X(t)}(\sigma)$.

The expectation of $X(t)$ (the average value) is defined as

$$\langle X(t) \rangle = \int_{\gamma=-\infty}^{+\infty} \gamma p_{X(t)}(\gamma)d\gamma.$$

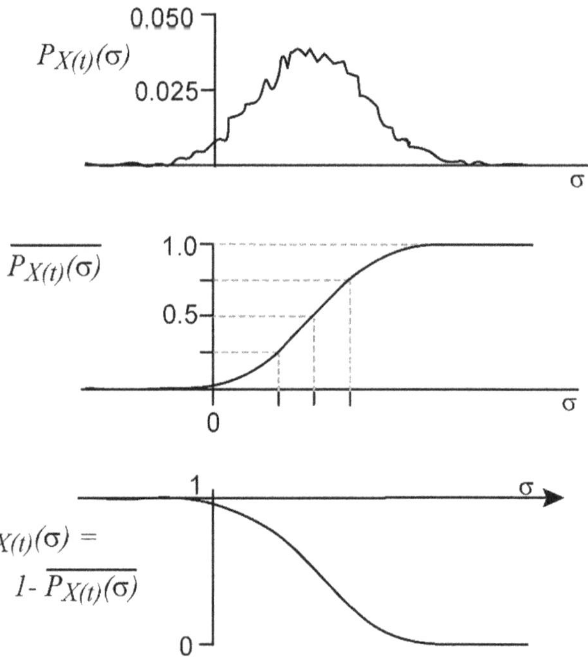

Figure 2.7. Complements of the distribution of first latency for Na$^+$ channel openings, schematic representation. The bi-exponential fit suggests multiple closed states and slow activation kinetics comparable to macroscopic inactivation.

2.5.2 Lifetimes in a single state with one exit pathway

Consider a single open state (O) with one exit pathway whose rate constant is a:

$$A \underset{a}{\rightleftharpoons} O.$$

Let t_0 be a random variable equal to the lifetime of a sojourn in O. What is mandated by the Markovian assumption is that $P_{t_o}(t)$, the probability that $t_o > t$, is given by an exponential function of t. $P_{t_o}(t)$ is otherwise known as the complement of the distribution for t_0. What is meant by the diagram is that

$$P_{t_o}(t) = \exp{(-at)}.$$

This is equivalent to

$$\frac{dP_{t_o}(t)}{dt} = -aP_{t_o}(t) \text{ or } A \xleftarrow{a} O, \tag{2.3}$$

where A is an 'absorbing state,' i.e. one from which there is no escape.

Two corollary features follow from the above three equivalent expressions for the lifetime of a sojourn in O. First, the probability density function for t_o, which we will call $p_{t_o}(t)$, is given by $-dP_{t_o}/dt = a \exp{(-at)}$. It follows that the mean lifetime of a sojourn in O is

$$<t_o> = \int_{\sigma=0}^{\infty} \sigma a \exp{(-a\sigma)}d\sigma = 1/a.$$

Second, equation (2.3) points to an intuitive meaning for the rate constant a. The probability 'lost' through transition a in a small increment of time Δt centered at t is given by

$$-\Delta P_{t_o}(t) \sim -\frac{dP_{t_o}(t)}{dt}\Delta t = aP_{t_o}(t)\Delta t.$$

Hence, recalling that $\Pr\{A \cap B\} = \Pr\{A \mid B\}\Pr\{B\}$, $a\Delta t$ is close to the $\Pr\{$ exit through transition a in $[t, t + \Delta t]|$ in state O at $t\}$. Taking $\Delta t \to 0$, a is therefore exactly equal to the rate of probability loss through transition a, given that the channel is in state O. This is the 'meaning' of a rate constant. That is why rate constants have units of reciprocal time.

2.5.3 Lifetimes in a single state with multiple exit pathways

Consider a single open state (O) with two exit pathways whose rate constants are a and b:

$$A \underset{a}{\rightleftharpoons} O \overset{b}{\rightleftharpoons} B. \tag{2.4}$$

Let t_0 be a random variable equal to the lifetime of a sojourn in O. What is mandated by the Markovian assumption is that $P_{t_o}(t)$, the probability that $t_o > t$, is given by an exponential function of t. $P_{t_o}(t)$ is otherwise known as the complement of the distribution for t_0. What is meant by the diagram is that

$$P_{t_o}(t) = \exp\left(-(a + b)t\right).$$

This is equivalent to

$$\frac{dP_{t_o}(t)}{dt} = -(a + b)P_{t_o}(t). \tag{2.5}$$

The generalization to an arbitrary number of exit pathways is obvious: $P_{t_o}(t) = \exp\left(-(a + b + \ldots)t\right)$. Given this generalized form for the exponential lifetime in any individual state, it can be shown that the overall equations for the system of states would be like those shown in equation (2.2).

Three corollary features follow from the above two equivalent expressions for the lifetime of a sojourn in O. First, the probability density function for t_o, which we will call $p_{t_o}(t)$, is given by $-dP_{t_o}/dt = (a + b)\exp\left(-(a + b)t\right)$. It follows that the mean lifetime of a sojourn in O is

$$<t_o> = \int_{\sigma=0}^{\infty} \sigma(a + b)\exp\left(-(a + b)\sigma\right)d\sigma = 1/(a + b).$$

For an arbitrary number of exit pathways, $\langle t_o \rangle = 1/(a + b + \ldots)$.

Second, from the form of expressions like those in equation (2.2), it is clear that the diagram in equation (2.4) specifies a precise bias for exit from O to A vs. B. The rules for exiting from O to A or B must be specified by parameters associated only with O (i.e. a, b), otherwise we would violate the Markovian assumption. So, in terms of calculating whether the exit will be to A or B, it does not matter what the rate constants are for $B \rightarrow O$ or $A \rightarrow O$ (see Box 2.2). We might as well set them to zero, yielding:

$$A \xleftarrow[a]{} O \xrightarrow{b} B. \tag{2.6}$$

Box 2.2. The Chinese banquet hypothesis—analogy for lifetimes with two exit pathways.

... the best Chinese restaurant in Baltimore, Szechuan House, not saying it is a good Chinese restaurant but the best Chinese restaurant in Baltimore. And one of their best dishes is Kung Pao chicken. So imagine we are seated at a table with a plate of Kung Pao chicken sizzling—I love it, can't wait to bite into that with the hot peppers, sometimes in the mood, I eat a hot pepper. And, that is wonderful. Now, imagine there is a barrier between two people. This barrier doesn't prevent you from eating from the plate, but it inhibits you from seeing anyone else eating from it. So in front of this plate, one of them, we have this very slender well-mannered soft spoken young woman, she is at the meal picking at the plate with a lone pair of chopsticks. On the other side, we have this gentleman, massive, big, and hungry. He has got a pair of chopsticks too. And can use it very well. They both eat from this plate, but they can't see each other. She doesn't know there is this big guy, and this big guy doesn't know there is this petite girl. So what we do is we calculate the average lifetime of a

> *piece of Kung Pao chicken disappears going one way versus another. Our intuition is, well, the lifetime going of the piece of chicken going [towards the big guy] is a lot shorter than towards [the petite girl]. It is obvious. Look at the relative size of these individuals. But in actuality, the average lifetime of a piece of Kung Pao chicken that goes towards the petite girl is exactly the same as the average lifetime of the piece that goes towards the big guy. The answer to this paradox—is what we will cover in next lecture.*
>
> *—David Yue*

Applying expressions of the form in equation (2.2) (chemical reaction style) to equation (2.3), we can write that

$$\frac{dA}{dt} = aO = a \exp\left(-(a+b)t\right) \tag{2.7a}$$

$$\frac{dB}{dt} = bO = b \exp\left(-(a+b)t\right). \tag{2.7b}$$

The overall probability that a channel in O will ever exit to A or B (the 'ever' probability) is then given by integrating the above equations over all t, yielding

$$\text{ever } \Pr\{O \to A\} = \int_{\sigma=0}^{\infty} a \exp\left(-(a+b)\sigma\right)d\sigma = a/(a+b)$$

$$\text{ever } \Pr\{O \to B\} = \int_{\sigma=0}^{\infty} b \exp\left(-(a+b)\sigma\right)d\sigma = b/(a+b).$$

The $\Pr\{O \to A \mid O \cap \text{exit}\}$ and $\Pr\{O \to B \mid O \cap \text{exit}\}$ in *any* infinitesimal time interval $[t + \Delta t]$ are equivalent to *ever* $\Pr\{O \to A\}$ and *ever* $\Pr\{O \to B\}$, respectively, by the Markovian assumption.

Third, equation (2.7) points to an intuitive meaning for rate constants a and b. The probability 'lost' through transition a in a small increment of time Δt centered at t is given by

$$\Delta A(t) \sim \frac{dA(t)}{dt}\Delta t = aO\Delta t.$$

Hence, recalling that $\Pr\{A \cap B\} = \Pr\{A \mid B\}\Pr\{B\}$, it follows that $a\Delta t$ is very close to the $\Pr\{$ exit through transition a in $[t, t + \Delta t]|$ in state O at $t\}$. Taking $\Delta t \to 0$, a is therefore exactly equal to the rate of probability loss through transition a, given that the channel is in state O. Likewise, b is the rate of probability loss through transition b, given that the channel is in state O. This generalizes the 'meaning' of a rate constant to states with multiple exit paths.

2.5.4 Binomial statistics

When there are n identical and independent units, each of which can either be ON or OFF (this is the binomial part of it), the probability of observing m units ON in a given observation (define it as $P(m)$) comes from binomial statistics. Let p be the probability that a unit will be ON in a given trial. The probability of a unit being off is therefore $q = 1 - p$. We will show that

$$P(m) = \frac{n!}{(n-m)!m!} p^m (1-p)^{n-m}.$$

Here, $p^m(1-p)^{n-m}$ is the probability that one particular set of m units is ON (the other $n-m$ units are OFF). But all units are identical, so we are not interested in the probability that one particular set of m units is ON. Rather, we want the probability that any set of m units is ON. So we have to multiply this initial entity by the number of ways of choosing m units out of a total of n units. Imagine each of n units is painted with a unique number from one to n. Now we are going to pick out m units, one after the other. How many ways are there of doing this? On the first 'draw,' we can pick any one of n units. On the second 'draw,' we can pick out any one of $n-1$ units. And so forth, until we have undertaken the mth draw.

There are

$$n(n-1)(n-2) \dots (n-m+1) = \frac{n!}{(n-m)!}$$

ways of drawing out m units if the order in which we draw the units is important. For example, if there are eight units and we are drawing out four units, $\{8, 6, 1, 2\}$ is counted as a different way of drawing out m units than $\{1, 2, 8, 6\}$. But we do not care about the 'order' we pick out a combination of m units. 'Order' is an epiphenomenon of the way in which we have conceptualized the process. $n!/(n-m)!$ overcounts the number of ways in which m objects can be chosen from n units by the number of ways in which m units can be ordered. There are $m!$ ways of ordering m units. (The logic for this is as follows. On the first of m draws, there are m ways to do this. On the second draw, there are $m-1$ ways. By the mth draw, there is only one way. Overall, there are $m!$ ways of selecting m units if order counts. Hence, the number of ways m units can be selected out of a total of n units is $n!/(n-m)!/m!$. This gives the correct multiplier for the probability terms. This entity is also known as a 'binomial coefficient' or 'n choose m' (also written $\binom{n}{m}$).

2.6 Single neuroblastoma Na channels

2.6.1 Elements of single-channel data

Figure 1 of Aldrich *et al* (1983) shows seven specimen records of single-channel activity obtained in the cell-attached mode from a neuroblastoma cell. Downward deflections represent openings; stacked openings are from the simultaneous activity of two channels. At the bottom is the average of many traces, divided by N, the estimated number of channels in the patch, and by i, the estimated size of an open channel. These give the waveform of open probability $P_O(t)$. This calculation is based on the assumption of independent and identical channels. Under this condition, the average current from a single channel is $\langle i(t) \rangle \equiv P_O(t)i$. If the N channels are independent, then $\langle i_1(t) + \cdots + i_N(t) \rangle = \langle i_1(t) \rangle + \cdots + \langle i_N(t) \rangle$. If the channels are identical, then the average current response of N channels is

$$\langle I(t) \rangle = N\langle i(t) \rangle = NP_O(t)i.$$

The $P_O(t)$ waveform has the expected characteristic, with a slow time constant of decay called τ_h, which, according to HH analysis, corresponds to the inactivation process.

Overall, the behavior of the $P_O(t)$ waveform reproduces the expected behavior for a self-respecting Na channel, as the plots in figure 2 of Aldrich *et al* (1983) demonstrate. Thus, we are justified in looking at the single-channel details to glean insight into genuine Na channel gating.

From the single-channel data in figure 1 of Aldrich *et al* (1983), three measurements are made in the single-channel records: (1) the time until first opening L (otherwise known as the 'first latency'); (2) the duration of an opening D; and (3) the number of openings k in a sweep (shown to the right of the traces). These are random variables, the probability functions of which allow us to argue in favor of an actual gating mechanism like that in panel (c) of figure 2.6. Visual inspection already hints that such a mechanism is in effect.

2.6.2 Complement of the distribution of first latency argues that components of activation are as slow as macroscopic inactivation (τ_h)

Figure 3 of Aldrich *et al* (1983) shows the complement of the distribution of first latency L. In this paper, it is called $F(t)$, which corresponds to Pr{channel first opens at time $>t$}. The experimental $F(t)$ can be fitted with a function of the form:

$$F(t) = B + \frac{(1 - B)}{R_1 - R_2}(R_1 e^{-R_2 t} - R_2 e^{-R_1 t}),$$

where $R_1 > R_2$. Note the correction to the error in the manuscript. The first latency probability density function for first latency $f(t) = -\frac{dF(t)}{dt}$. The plateau value is B; this represents the fraction of trials in which a channel passes directly into the inactivated state (I) from the closed, resting state (C).

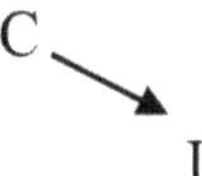

The bi-exponential form of F is consistent with a gating model in which two closed states line the pathway of activation. We will go into this further in the next chapter. For now, the important thing to note is that there are time constants in the fit to first latency that are slow enough to account for macroscopic inactivation ($1/R_2 \sim \tau_h$). This is surprising when considered in the context of the HH framework, since activation (which is gauged by F) is supposed to be very fast compared with inactivation. This provides a first hint that activation, as gauged by the first latency function, may be rate limiting, not genuine inactivation.

Note that in the case of multichannel patches, we initially calculate an apparent first latency distribution $\hat{F}(t)$, which is the Pr{time to the first opening of N channels (L) $> t$} Because all N channels must not have exhibited a first opening for L to continue,

$$F(t)^N = \hat{F}(t).$$

Hence, we can calculate $F(t)$ from the Nth root of $\hat{F}(t)$. Such 'compensated' functions are shown in figure 3 of Aldrich *et al* (1983).

2.6.3 Direct calculation of the inactivation rate from the open state

To make the claim that activation is rate limiting and inactivation is rapid, we must do more than simply show slow components in $F(t)$. We must directly determine the rate of inactivation from the open state and show that it is fast relative to the macroscopic inactivation of the current. Aldrich and colleagues (Aldrich *et al* 1983) found that the complement of the distribution for open durations D defined a single-exponential decline, arguing for a single open state O (from the reasoning in section 2.3.2). Specifically, then, what we need to do is deduce the rate constant b and show that it is $\gg 1/\tau_h$.

$$\begin{array}{c} \xleftarrow{a} O \\ b\downarrow \\ I \end{array}$$

Inspection of the mean open times (figure 5, bottom panel (Aldrich *et al* 1983)) already indicates that open times last about 0.5 ms, ($\sim\tau_o$). From section 2.5.3, we know that $a + b \sim 2$ m s^{-1}. If the deactivation rate a were small, then we could conclude that b is very fast relative to macroscopic inactivation. But we need to prove this.

We need another constraint on $a + b$. To obtain this, we consider the statistics for the number of openings per sweep k. These can be predicted by viewing the gating model as a discrete Markov chain, linked by 'ever' probabilities (section 2.5.3).

$$\begin{array}{ccc} & A & \\ C & \rightleftarrows & O \\ & B & \\ 1\text{-}A \searrow & & \swarrow 1\text{-}B \\ & I & \end{array}$$

Here, A is the probability that a channel in C will transition to O rather than I in the next transition, while B is the probability that a channel in O will transition to C rather than I in the next transition. A can be determined from the data as follows:

$$1 - A = (\text{fraction of blank sweeps})^{1/N}.$$

From section 2.5.3, we know that B is related to a and b by

$$B = \frac{a}{a + b} = a\tau_0.$$

Hence, if we could specify B, then we could individually specify a and b. By considering the Markov chain above, we can derive an expression for $P(k)$, the probability of observing k openings in a given sweep, based on the values of A and

B. Because we can directly calculate A, fits of $P(k)$ to data could yield B, thereby enabling us to calculate b:

$$P(k) = \left(\frac{1 - AB}{B}\right)(AB)^k ; k \geqslant 1$$
$$P(0) = 1 - A.$$

This equation can be derived as follows. For no openings, $P(0) = 1 - A$, by definition. By tracing the possible trajectories leading to k openings, the following expressions are derived, under the assumption that the transition probabilities only care about the state of residence at a given instant of time:

$$P(0) = 1 - A$$
$$P(1) = A[(1 - B) + B(1 - A)] = A(1 - AB)$$
$$P(2) = ABA[(1 - B) + B(1 - A)] = ABA(1 - AB)$$
$$\vdots$$
$$P(k) = (AB)^{k-1}A[(1 - B) + B(1 - A)] = A^k B^{k-1}(1 - AB) = \left(\frac{1 - AB}{B}\right)(AB)^k.$$

When there is more than one channel, the apparent $P(k)_N$ is given by N-fold convolution of $P(k)$, as follows. This follows from a general property of the sum of independent random variables. The probability density function for $\underbrace{(X + X + \cdots X)}_{N \text{ terms}} = Y$ is given by the convolution of the probability density functions for the individual terms. We can essentially prove this by following the logic for two channels:

$$P(0)_2 = P(0)P(0) = (1 - A)^2$$
$$P(1)_2 = P(0)P(1) + P(1)P(0)$$
$$P(2)_2 = P(0)P(2) + P(1)P(1) + P(2)P(0)$$
$$\vdots$$
$$P(k)_2 = \sum_{X_1=0}^{k} P(X_1)P(k - X_1).$$

The latter expression is a twofold discrete convolution. For N channels, we have the N-fold convolution:

$$P(k)_N = \sum_{X_1=0}^{k} \sum_{X_2=0}^{k-X_1} \sum_{X_3=0}^{k-X_1-X_2} \cdots \sum_{X_{N-1}=0}^{k-X_1-\ldots-X_{N-2}} P(X_1) \ldots P(X_{N-1})P(k - X_1 - \ldots - X_{N-1}).$$

These convolutions tally all the ways in which channels can produce k openings on a given sweep. Fitting such equations to the data, as shown in figure 4 (Aldrich *et al* 1983), yields the actual values of B as well as A. B turns out to be about 0.2 (see the top of figure 5 in Aldrich *et al* 1983). Hence, most channels open once, then inactivate to I. In other words, $b \gg a$. Hence,

$$b = \frac{1 - B}{\tau_o} = \frac{0.8}{0.5} = 1.6 \text{ ms}^{-1}.$$

This corresponds to characteristic microscopic inactivation times of about two-thirds of a millisecond, far faster than the macroscopic inactivation represented by τ_h. The bottom of figure 5 in Aldrich *et al* (1983) demonstrates the point graphically. Note at the bottom of figure 5 that the mean open duration is given by $1/(a + b) \sim 1/b$. If microscopic inactivation accounted for macroscopic inactivation, the data points should fall on the line of identity.

2.6.4 Convolution analysis allows a graphical assessment of the contributors to macroscopic inactivation

An alternate way to view the operation of the channel gating system is to recast it in terms of a classic input/output relation for a time-invariant, linear system:

$$f(t) \longrightarrow [M(\tau)] \longrightarrow P_O(t),$$

where $f(t) = -dF/dt =$ the rate of first opening probability, $M(\tau)$ is $\text{Pr}\{$ in state O at $t + \tau \mid$ in state O at $t\}$, and $P_O(t)$ is $\text{Pr}\{$ in state O at $t\}$. By analogy to a linear system, the 'output' $P_O(t)$ is given by the convolution of the 'input' $f(t)$ and the 'impulse response' of the system $M(\tau)$, specifically:

$$P_O(t) = \int_{\tau=0}^{t} f(t - \tau)M(\tau)d\tau.$$

Independent of the analogy to linear systems, we can justify this expression by reasoning from probability. If a channel is found to be open at time t, it must have successfully undergone two independent events. It must have first opened in a small Δt near $t - \tau$ (event $\mathcal{A}(t - \tau)$), and it must be open after a delay τ following the first event (event $\mathcal{B}(\tau)$). The two events are independent, according to the Markovian assumption. The probability of these two events happening is

$$\text{Pr}\{\mathcal{A}(t - \tau) \cap \mathcal{B}(\tau)\} = \text{Pr}\{\mathcal{A}(t - \tau)\}\text{Pr}\{\mathcal{B}(\tau)\mid \mathcal{A}(t - \tau)\}$$
$$= f(t - \tau) \cdot \Delta t \cdot M(\tau).$$

The total number of ways of finding a channel open at t can be tallied by considering all instances of $\mathcal{A} \cap \mathcal{B}$ for all τ. The overall probability of being open is then obtained by summing the probabilities of all cases of $\mathcal{A} \cap \mathcal{B}$. As $\Delta t \to 0$, the sum of all such probabilities (in the expression above) evaluated over all possible $\tau[0, t]$ becomes the convolution expression. A more rigorous derivation, based on systems theory, is included later as an addendum for the curious reader.

We can now turn to the graphical insight, which is shown in figure 6 of Aldrich *et al* (1983). There are three convolutions here, one corresponding to a different step voltage. Let us consider the $V = 10$ experiment for concreteness. In each convolution experiment, the top trace is $L(\tau)$, which is our $P_{t_o}(\tau)$ from section 2.5.3. This is the complement of the distribution for open durations ($\text{Pr}\{$opening lasts $> \tau\}$). We should be using $M(\tau)$ in the strict convolution experiment, but we are going to see whether $L(\tau) \sim M(\tau)$. This approximation means that we expect the channel to open only once. The proof of the pudding will be to see whether the convolution of $f(t)$ with $L(\tau)$ accurately predicts $P_O(t)$. The second trace in each convolution experiment is $f(t)$. The convolution of the top two traces (f and M) yields the smooth curve in the third row. The fact that this overlays the noisy experimental estimate of $P_O(t)$ (obtained as in figure 1) shows the adequacy of the approximation that $L(\tau) \sim M(\tau)$; hence, for all reasonable purposes, the channel only opens once, if at all, per sweep.

This leads us to three key points: first, one can also see that most of the slow time course of $P_O(t)$ comes from $f(t)$, not the shorter-lived $L(\tau)$ in the first row. This graphical interrelation holds for the other voltages.

Second, there is a useful analogy to low-pass filtering. $L(\tau)$ is a single exponential. The impulse response of a first-order low-pass filter is also a single exponential. Because low-pass filtering is equivalent to convolving an initial signal with the impulse response of a filter, $P_O(t)$ can be considered a low-pass-filtered version of $f(t)$. If $L(\tau)$ is very short-lived compared to $f(t)$, then $L(\tau)/\tau$ could be considered a 'delta function,' whose properties are such that $P_O(t) = f(t)\tau$.

Third, breaking up gating into a function based on activation $f(t)$ and subsequent gating $M(\tau)$ simplifies the analysis of channel gating, as one can focus on one subproblem at a time.

References

Aldrich R W 1986 Voltage-dependent gating of sodium channels: towards an integrated approach *Trends Neurosci.* **9** 82–6

Aldrich R W, Corey D P and Stevens C F 1983 A reinterpretation of mammalian sodium channel gating based on single channel recording *Nature* **306** 436–41

Eyring H 1935 The activated complex in chemical reactions *J. Chem. Phys.* **3** 107–15

Hille B 2001 *Ion Channels of Excitable Membranes* 3rd edn (Sunderland, MA: Sinauer)

Hodgkin A L and Huxley A F 1952 A quantitative description of membrane current and its application to conduction and excitation in nerve *J. Physiol.* **117** 500–44

Yue D T, Lawrence J H and Marban E 1989 Two molecular transitions influence cardiac sodium channel gating *Science* **244** 349–52

IOP Publishing

Ion Channel Gating and Mechanisms

David T Yue, Manu Ben-Johny and Ivy E Dick

Chapter 3

Markovian gating as a linear system I

3.1 Introduction

The paper by R W Aldrich, D P Corey, and C F Stevens (ACS) (Aldrich *et al* 1983) illustrated just how important it is to understand the stochastic basis of single-channel information. A type of stochastic process, called a Markov process, was supported by experimental measures of open durations in several types of channels. Additionally, simple reasoning based on Markov mechanisms was essential to interpreting single-channel data. In this chapter, we introduce the theory of Markov processes in a more complete, systematic manner. We show that it has clear links to linear system theory, which provides us with clear insights into the workings of Markov processes.

The first half of the chapter is devoted to understanding 'discrete-step' Markov chains. Here the 'clock' is not time, but an integer transition number. In particular, we ask which state a channel will occupy after k transitions, regardless of when the transitions occur. This is a 'full-blown' theory involving 'ever' probabilities.

The second half of the unit is devoted to continuous-time Markov processes. Here, we calculate the probabilities of occupying various states as a function of continuous time.

3.2 Discrete-step Markov chains

3.2.1 Definition

A discrete-step Markov chain is a stochastic process related to the set of random variables $X(k)$, where k is a nonnegative integer referring to the number of state transitions a system has undergone since some initial observation time, and $X(k)$ is assigned a nonnegative integer depending upon the number of the state occupied after k transitions. The Markovian ('memoryless') assumption is equivalent to the assumption that

doi:10.1088/978-0-7503-2388-8ch3　　3-1　　

$$\pi_{ij}(k) \equiv \text{ever Pr\{transition from state } i \text{ to } j \text{ on the } k\text{th transition}|$$
$$\text{in state } i \text{ after } (k-1) \text{ transitions}\}$$

is equal to a constant π_{ij}, regardless of the value of k and the sequence of transitions before transition k.

3.2.2 Stochastic matrix representation of Markov chains

Consider the simple gating model below, in which we only consider where the channel resides after k transitions (staying put in a state does not count as a transition):

$$C_1 \underset{1-A}{\overset{1}{\rightleftharpoons}} C_2 \underset{1}{\overset{A}{\leftrightharpoons}} O_3,$$

where the values of the 'ever' probabilities are drawn next to the appropriate arrows. Inspection tells us that we could express the transition rules for this system as the following matrix equation:

$$\left[P_{C_1}(k+1) \; P_{C_2}(k+1) \; P_{O_3}(k+1) \right] =$$
$$\left[P_{C_1}(k) \; P_{C_2}(k) \; P_{O_3}(k) \right] \begin{bmatrix} 0 & 1 & 0 \\ 1-A & 0 & A \\ 0 & 1 & 0 \end{bmatrix}.$$

It is clear from this example that this sort of matrix equation can be generalized to Markov chain mechanisms with an arbitrary number of states n such that

$$\left[\vec{P}(k+1) \right] = \left[\vec{P}(k) \right][\pi_{ij}], \tag{3.1}$$

where $\vec{P}$ is a probability row vector ($1 \times n$) and π_{ij} are appropriate 'ever' Pr contained in an $n \times n$ matrix, π. In this formulation, i and j also give the row and column, respectively, in which an 'ever' Pr is found. We can make two remarks about the 'master equation' for Markov chains:
  (1) The condition of the system is no longer represented by a deterministic 'state vector' (as would be the case in a linear system) but by an analogous probability row vector $P(k)$.
  (2) The matrix containing all the π_{ij}, which we refer to as π, is a 'stochastic matrix' because the sum of elements in any row is equal to unity. This follows from the fact that the sum of elements in any row is given by $\sum_{j} \pi_{ij}$, which is the probability that the channel will be in some state after the next transition, given that it was in state i after the current number of transitions.

The solution to the general matrix equation defining a Markov chain is self-evident from straightforward manipulation of the master equation:

$$P(k + 1) = P(k)[\pi]$$
$$P(k + 2) = P(k + 1)[\pi] = P(k)[\pi]^2$$
$$P(k + 3) = P(k + 2)[\pi] = P(k)[\pi]^3 \qquad (3.2)$$
$$\vdots$$
$$P(k + l) = P(k)[\pi]^l.$$

Equation (3.2) is the discrete-step Chapman–Kolmogorov equation. Hence, a stochastic process that obeys the discrete-step Chapman–Kolmogorov equation is equivalent to a discrete-step stochastic process that obeys the memoryless assumption in section 3.2.1.

This Chapman–Kolmogorov equation is identical to the homogeneous part of a discrete-time state-variable equation from linear systems. Hence, all the techniques of z-transformation that are so helpful in the latter type of equation can be applied to Markov chains. We will not develop this in this chapter but mention it as an obvious fund of resources that can be applied to Markov chains. An interesting application of the discrete Chapman–Kolmogorov equation to ion channel modulation may be found in Herzig $et\ al$ (1993), 'Mechanisms of β-adrenergic stimulation of cardiac Ca^{2+} channels revealed by discrete-time Markov analysis of slow gating.'

3.2.3 Canonical master equation for Markov chains

All states belong to either the 'absorbing' class $\mathcal{A}$ or the 'transient' class $\mathcal{T}$. There is no exit from the $\mathcal{A}$ class, but it is possible to exit the $\mathcal{T}$ class of states (Box 3.1).

Example 1. *An example of this classification of states comes from the analysis by ACS in Aldrich et al (1983) (figure 3.1). Here, $\mathcal{T} = \{C, O\}$ and $\mathcal{A} = \{I\}$.*

Box 3.1. The Roach Motel—analogy for absorbing and transient states.

There are these little traps for roaches called the Roach Motel. Roaches can go into these traps and there is some kind of a sticky paper that prevents the roaches from getting out once they get in. So the way they advertised it was— 'they check in but they don't check out.' That is how you can think of these absorbing states—you can check in but you can't check out. The transient states are outside of the roach motel

—David Yue

Example 2. *In calculating whether a channel that starts out in C will inactivate before ever opening, it is useful to consider certain states to be in $\mathcal{A}$, even though they are not literally part of the absorbing class of states. For example, we may find it useful to consider $\mathcal{T} = \{C\}$ and $\mathcal{A} = \{O, I\}$ (figure 3.1).*

Because of the Markovian assumption, the exit probabilities of first leaving C are unchanged from the original system. Exit probabilities from C can only be a function of parameters associated with C. The advantage of the recast system is that the

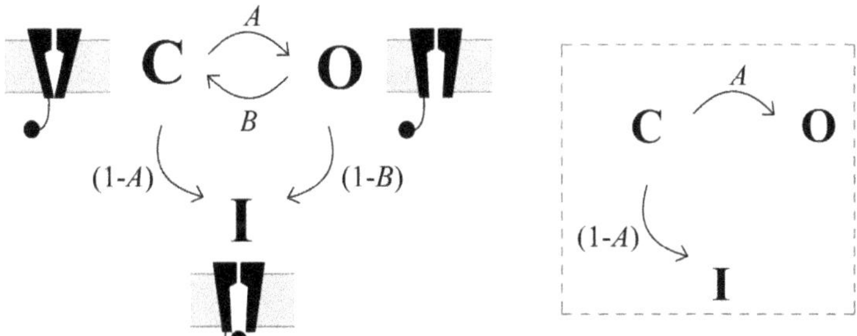

Figure 3.1. State-transition diagram from the ACS model (Aldrich *et al* 1983). The closed (*C*) and open (*O*) states are considered transient states, while the inactive state (*I*) is considered an absorbing state, given the 'slow' time course for recovery from inactivation. In some cases, it may be useful to consider certain states as absorbing to facilitate the calculation of certain parameters. For example, {*O, I*} can be treated as absorbing to calculate the likelihood of a channel inactivating before opening (right inset).

probability of ending up in O now conveniently tallies the probability of opening at least once before inactivating, and the probability of ending up in I now conveniently tallies the probability of inactivating before ever opening.

This classification of states lends itself to a 'canonical' master equation for Markov chains:

$$[\boldsymbol{P}_T(k+1) \vdots \boldsymbol{P}_A(k+1)] = [\boldsymbol{P}_T(k) \vdots \boldsymbol{P}_A(k)] \begin{bmatrix} \pi_{TT} & \vdots & \pi_{TA} \\ \cdots & \vdots & \cdots \\ \pi_{AT} & \vdots & \pi_{AA} \end{bmatrix}.$$

The advantage of the canonical form is that much of the behavior of the system can be predicted by only considering a portion of the overall π matrix. For one thing, $\pi_{AT} = 0$ because there is no exit from the absorbing states. Second, it is evident that π_{TT} completely specifies the behavior of states in the transient class because

$$[\boldsymbol{P}_T(k+1)] = [\boldsymbol{P}_T(k)][\pi_{TT}],$$

so that

$$[\boldsymbol{P}_T(l)] = [\boldsymbol{P}_T(0)][\pi_{TT}]^l.$$

An interesting property of π_{TT} is that $\lim_{m\to\infty} \pi_{TT}^m = 0$ because $\boldsymbol{P}_T(\infty) = 0$ by definition.

Example 3. *For the full ACS system (Aldrich et al 1983), the canonical matrix π is*

$$\begin{bmatrix} \pi_{TT} & \vdots & \pi_{TA} \\ \cdots & \vdots & \cdots \\ \pi_{AT} & \vdots & \pi_{AA} \end{bmatrix} = \begin{bmatrix} 0 & A & \vdots & 1-A \\ B & 0 & & 1-B \\ \cdots & & \vdots & \cdots \\ 0 & & \vdots & (nd) \end{bmatrix},$$

where

$$[\boldsymbol{P}_T(k)\!:\!\boldsymbol{P}_A(k)] = [P_C(k)P_O(k)\!:\!P_X(k)].$$

3.2.4 Fundamental matrix M simplifies calculation of behavior in $\mathcal{T}$

3.2.4.1 Expected number of visits to a transient state before absorption

Because π_{TT} completely determines behavior in $\mathcal{T}$, it is not surprising that it can be used to define another matrix that directly yields many properties of the transient class. This other matrix is known as the fundamental matrix M (Colquhoun and Hawkes 1981), which obviates laborious, ad hoc calculations that have so far (in the ACS paper (Aldrich *et al* 1983)) been required to specify certain key properties of the transient class.

$$M \equiv [I - \pi_{TT}]^{-1} \underset{\text{by laws of algebra}}{\equiv} I + \pi_{TT} + \pi_{TT}^2 + \ldots$$

The second equality can be seen from 'long division' so long as the sequence converges.

$$
I - \pi_{TT} \overline{\smash{\big)}\ I}
$$

$$
\begin{array}{r}
I + \pi_{TT} + \pi_{TT}^2 + \ldots \\[4pt]
I \\[2pt]
\underline{I - \pi_{TT}} \\[2pt]
\pi_{TT} \\[2pt]
\underline{\pi_{TT} - \pi_{TT}^2} \\[2pt]
\pi_{TT}^2 \\[2pt]
\pi_{TT}^2 - \pi_{TT}^3
\end{array}
$$

From the series definition of M, we can see that the ij element is equivalent to

$$M_{ij} = \delta_{ij} + [\pi_{TT}]_{ij} + [\pi_{TT}]_{ij}^2 + \ldots,$$

where each of the terms is Pr {in state j after k transitions |starting in state i at $k = 0$} with $k = 0$, 1, 2, ..., respectively. To understand the meaning of M_{ij}, define a random variable $X_{ij}(k)$, which is set to zero if the channel is not in state j after k transitions or set to one if the channel is in state j after k transitions. This assumes that the channel starts in state i before any transitions occur ($k = 0$). $X_{ij}(k)$ therefore tallies a visit to state j after k transitions, given a start in state i. The expected value of $X_{ij}(k)$, accumulated from tallies over all possible numbers of transitions (k), is therefore given by

$$
\langle X_{ij}(k)\rangle_k = \sum_{k=0}^{\infty} \left\{ \begin{array}{l} 0 \times (1 - \text{Pr}\{\text{in state } j \text{ after } k \text{ transitions}\}) + \\ 1 \times \text{Pr}\{\text{in state } j \text{ after } k \text{ transitions}\} \end{array} \right\}
$$

$$
= \sum_{k=0}^{\infty} \text{Pr}\{\text{in state } j \text{ after } k \text{ transitions} \mid \text{in state } i \text{ on } k = 0\}
$$

$$
= M_{ij}.
$$

Hence, M_{ij} is the expected number of visits to state j before the system enters an absorbing state $\mathcal{A}$, given that the channel starts in state i.

Example 4. *For the ACS system in example 3, let us calculate the mean number of visits to the closed states before the channel is inactivated, given that the channel starts the depolarizing epoch in C. We have that*

$$\pi_{TT} = \begin{bmatrix} 0 & A \\ B & 0 \end{bmatrix}.$$

Hence, M is

$$\boldsymbol{M} = [\boldsymbol{I} - \pi_{TT}]^{-1} = \begin{bmatrix} 1 & -A \\ -B & 1 \end{bmatrix}^{-1} = \frac{1}{1 - AB} \begin{bmatrix} 1 & A \\ B & 1 \end{bmatrix},$$

and the desired element is the 1,1 element

$$\frac{1}{1 - AB}.$$

3.2.4.2 Multistep 'ever' probabilities from M

Another interesting property comes from $\boldsymbol{B} = \boldsymbol{M}\pi_{TA}$. It turns out that B_{ij} is the probability that a channel will first enter the absorbing set of states in state $j \in \mathcal{A}$, given that the channel starts off in state $i \in \mathcal{T}$. We can prove this, based on the Markovian assumption, by writing a recursive relation for B:

$$\boldsymbol{B} = \pi_{TA} + \pi_{TT}\boldsymbol{B}.$$

The ij elements of the first term give the probability of first being absorbed by state j after one transition. The second term gives the probability of first being absorbed by state j and not being absorbed after one transition. π_{TT} concerns the probability of still being in the transient set of classes after one transition, while B concerns the conditional probability of being first absorbed by state j, given that the channel was not absorbed after one transition (this is the same $\boldsymbol{B}$ because the Markovian assumption renders the added condition irrelevant). Solving for B yields

$$\boldsymbol{B} = [\boldsymbol{I} - \pi_{TT}]^{-1}\pi_{TA},$$

which is equivalent to our original definition for $\boldsymbol{B}$.

Example 5. *Consider the following channel gating model (figure 3.2):*

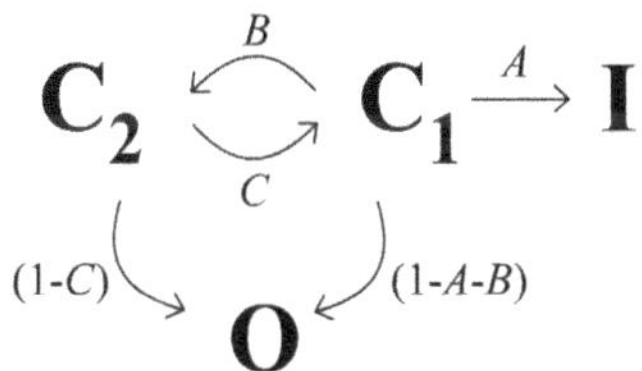

Figure 3.2. State transition diagram for example 5.

If we define $\boldsymbol{P} = [P_{C_1} P_{C_2} \vdots P_O P_I]$, we then have:

$$\pi = \begin{bmatrix} 0 & B & \vdots & A & 1 - A - B \\ C & 0 & \vdots & 1 - C & 0 \\ \cdots & \cdots & \vdots & \cdots & \cdots \\ 0 & 0 & \vdots & nd_- & nd \\ 0 & 0 & \vdots & nd & nd \end{bmatrix}$$

$$\begin{aligned}
\boldsymbol{B} &= [\mathbf{I} - \pi_{TT}]^{-1} \pi_{TA} \\
&= \begin{bmatrix} 1 & -B \\ -C & 1 \end{bmatrix}^{-1} \begin{bmatrix} A & 1 - A - B \\ 1 - C & 0 \end{bmatrix} \\
&= \frac{1}{1 - BC} \begin{bmatrix} 1 & B \\ C & 1 \end{bmatrix} \begin{bmatrix} A & 1 - A - B \\ 1 - C & 0 \end{bmatrix} \\
&= \frac{1}{1 - BC} \begin{bmatrix} A + B - BC & 1 - A - B \\ AC + 1 - C & C - CA - CB \end{bmatrix}.
\end{aligned}$$

So the probability of starting out in C_1 and first being absorbed by O is given by the 1,1 element, which equals $\frac{A + B - BC}{1 - BC}$.

3.2.5 Graph theory approach to Markov chains

It turns out that electrical engineers had already developed a graph theory method for calculating multistep 'ever' probabilities. This method is often much easier to calculate than the M matrix method described above. We introduce this to illustrate the enormous power of graph theory in probability analysis.

The usual context is to calculate the input–output gain of a system of black-box elements. Consider the following feedback system (figure 3.3).

The gain, defined by the ratio $\frac{Y}{X}$, is given by $\frac{G_1 G_2}{(1 - G_1 H)}$. Sam Mason developed the following graph theory rule (Mason 1953) for calculating the gain:

$$\frac{Y}{X} = \frac{1}{\Delta} \sum_{k=1}^{N} M_k \Delta_k,$$

where:

(1) M_k is the gain of the kth forward path, of which there are N;

(2) $\Delta = 1 - \underbrace{\sum_{m1} P_{m1}}_{\substack{\text{Gain product} \\ \text{of } mth \text{ loop}}} + \underbrace{\sum_{m2} P_{m2}}_{\substack{\text{Gain product} \\ \text{of } mth \\ \text{combination of} \\ \text{two non} - \text{touching} \\ \text{loops}}} - \underbrace{\sum_{m3} P_{m3}}_{\substack{\text{Gain product} \\ \text{of } mth \\ \text{combination of} \\ \text{three non} - \text{touching} \\ \text{loops}}} + \ldots;$

(3) Δ_k is the Δ for that part of the signal flow graph which does not touch the kth forward path.

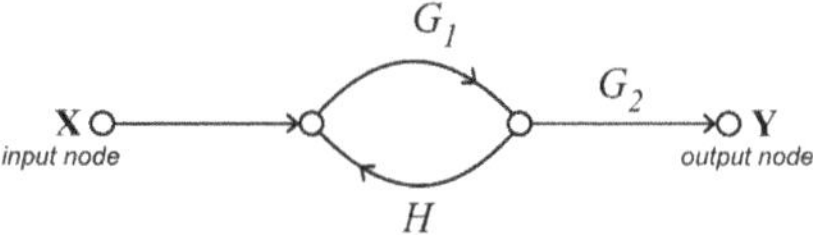

Figure 3.3. An example graph for a feedback system.

Remarks: 'touching' means sharing one or more nodes. A forward path cannot, in itself, make a loop.

The key insight is that this rule can also be applied to Markov chains to obtain the multistep 'ever' probabilities.

Example 6. *Solve the previous example by Mason's rule.*

There are two forward paths from C_1 to O, with gains

$$M_1 = A$$
$$M_2 = B(1 - C).$$

There are no non-touching loops for either forward path, so $\Delta_1 = \Delta_2 = 1$, and there is only one loop, so $\Delta = 1 - BC$. Plugging this into Mason's rule, the multistep 'ever' probability of transitioning from C to O is

$$\frac{A \times 1 + B(1 - C) \times 1}{1 - BC},$$

which is equivalent to the solution obtained by calculating matrix B in the previous example!

The equivalence of calculating multistep 'ever' probabilities by the B matrix method and Mason's rule raises an interesting philosophical viewpoint on the meaning of 'gain.' B_{ij} is the sum of the probabilities of all possible paths (Markov chains) that start in i and are first absorbed in j. The probability of any such path is equal to the product of the single-step 'ever' probabilities for each transition crossed in the path. Hence, the gain of a circuit is equivalent to the unweighted sum of the gain products of any possible path between input and output nodes. The unweighted feature means that any path between input and output nodes is equally likely. The fact that any possible path is included means that complex wanderings, including multiple loops, are reflected in the overall gain.

Explicitly, then, the output of a circuit Y as a function of its input X can be expressed as:

$$Y = (\text{gain product of pathway 1}) \times X +$$
$$(\text{gain product of pathway 2}) \times X +$$
$$\cdots$$

3.3 Continuous-time Markov processes

Given the submillisecond resolution of channel gating made possible by patch-clamp techniques, it is also necessary to understand the moment-to-moment changes in channel conformation in continuous time. Hence, we need to introduce continuous-time Markov processes.

3.3.1 Definition

A continuous-time Markov process is a stochastic process relating to the set of random variables $X(t)$, where t is a real number specifying continuous time, and $X(t)$ is assigned a nonnegative integer depending upon the number of the state occupied at time t. The key Markovian ('memoryless') feature of this process is this: for all $t \geqslant 0$ and $s \in \mathcal{R}$,

$$\Pr\{X(t+s) = j | X(s) = i \text{ and } X(u) = x(u), -\infty \leqslant u < s\} = \\ \Pr\{X(t+s) = j | X(s) = i\}.$$

In other words, the behavior of the future $X(t+s)$ depends only upon the present $X(s)$, independent of whatever the history of $X(u)$ has been.

3.3.2 Memory, exponential life, and the Chapman–Kolmogorov forward equations

What does the memoryless assumption imply about the kinetics of a system? What do exponentially distributed dwell times in states imply about memoryless behavior? What are the microscopic assumptions underlying our ability to write chemical reaction equations? Here, we will show that three properties—memoryless behavior, exponential lifetimes, and the Chapman–Kolmogorov forward equations (CKFEs), i.e. chemical reaction equations—are equivalent definitions for Markov processes. We can choose which definition to use according to what we need to show. The strength of the CKFEs is that they summarize Markovian behavior in terms of first-order differential equations, which, in some instances, are easier to apply than abstract definitional properties such as memoryless behavior. Therefore, what we need to show is:

$$\text{Memoryless behavior} \underset{\text{part I}}{\Longleftrightarrow} \text{Exponential dwell times} \underset{\text{part II}}{\Longleftrightarrow} \text{CKFEs.}$$

3.3.2.1 Part I: memoryless behavior $\Longleftrightarrow$ exponential dwell times in states
Here, we will show that memoryless behavior $\Rightarrow$ exponential dwell times. Let us consider D to be the duration of a sojourn in a single state which starts at time zero, i.e. D is the lifetime of the state. An equivalent definition for memoryless behavior is that, for any state,

$$\Pr\{D > s + T | D > T\} = \Pr\{D > s\},$$

where T and s are both $\geqslant 0$, T is some conditioning time at which we only choose openings that have remained open up to T, and s is the lifetime of the chosen openings beyond the conditioning time. The equivalent statement for memoryless behavior above indicates that states do not 'age.' In general, the exit from a state does not differ, regardless of whether it has been occupied for 1 msec or 1 year; all that matters is that we are in a certain state at a certain time. From the definition of a conditional probability, we can rewrite the above memoryless condition as

$$\frac{\Pr\{(D > s + T) \text{ and } (D > T)\}}{\Pr\{D > T\}} = \Pr\{D > s\},$$

which is equivalent to

$$\frac{\Pr\{D > s + T\}}{\Pr\{D > T\}} = \Pr\{D > s\}$$

or

$$\Pr\{D > s + T\} = \Pr\{D > T\}\Pr\{D > s\}. \tag{3.3}$$

Let $F(T)$ denote $\Pr\{D > T\}$. We can now calculate what the derivative of $F(t)$ must be:

$$\frac{dF}{dT} \equiv \lim_{s \to 0} \left\{ \frac{F(T + s) - F(T)}{s} \right\}.$$

Substituting from equation (3.3), we obtain:

$$\frac{dF}{dT} = \lim_{s \to 0} \left\{ \frac{F(T)F(s) - F(T)}{s} \right\}$$

$$\frac{dF}{dT} = F(T)\lim_{s \to 0} \left\{ \frac{F(s) - 1}{s} \right\}$$

$$\frac{dF}{dT} = F(T)\underbrace{\left(\frac{dF}{dT}(T = 0) \right)}_{-\alpha},$$

where α is a constant $\geqslant 0$. The solution to this differential equation must be a single exponential. Therefore, memoryless behavior $\Rightarrow$ single-exponential dwell times in any state.

To prove that single-exponential dwell times $\Rightarrow$ memoryless behavior, show that $F(T) = \exp(-\alpha T)$ satisfies equation (3.3). Applying the properties of an exponential function, we obtain: $\exp(-\alpha(s + T)) = \exp(-\alpha T)\exp(-\alpha s)$, satisfying equation (3.3). Equation (3.3) is equivalent to the memoryless assumption, as we have already shown.

3.3.2.2 Part II: exponential dwell times $\Leftrightarrow$ Chapman–Kolmogorov forward equations

First, we prove that exponential dwell times $\Rightarrow$ the CKFEs. Consider the channel to be in state i at time zero, while state j is any state that is one transition away from state i.

$$\text{Let } P_{ii}(h) = \Pr\{\text{in state } i \text{ at } h \mid \text{in state } i \text{ at time zero}\}$$

and

$$P_{ij}(h) = \Pr\{\text{in state } j \text{ at } h \mid \text{in state } i \text{ at time zero}\}$$

and

$$P_{ij} = \text{ever probability } i \to j.$$

It helps to first prove two lemmas.

Lemma 1. $1 - P_{ii}(h) = v_i h + o(h)$, where $o(h)$ is an order term which goes to zero faster than $h\,(\lim_{h\to 0} \frac{o(h)}{h} = 0)$, and v_i specifies the lifetime in state i as $\exp(-v_i t)$.

The proof relies on the assumption of an exponential dwell time in state i:

$$1 - P_{ii}(h) = \Pr\{\text{not in } i \text{ at } h\} = a(h) + b(h)$$

where

$$a(h) = \sum_j \Pr\{(\text{first } i \to j \text{ transition by } h) \text{ and (subsequently stayed in } j \text{ until } h)\}$$

and

$$b(h) = \sum_j \Pr\{(\text{first } i \to j \text{ transition by } h) \text{ and (subsequently left } j \text{ but not in } i \text{ at } h)\}.$$

Term a is given by

$$a(h) = \sum_j \int_{\sigma=0}^h v_i P_{ij} e^{-v_i \sigma}(e^{-v_j(h-\sigma)})d\sigma$$

$$= \sum_j v_i P_{ij}\left\{\frac{e^{-v_j h} - e^{-v_i h}}{v_i - v_j}\right\}$$

$$= \sum_j v_i P_{ij}\left\{\frac{1 - v_j h + o(h) - 1 + v_i h + o(h)}{v_i - v_j}\right\}$$

$$= v_i h \sum_j P_{ij} + o(h) = v_i h + o(h).$$

Term b is bounded by

$$b(h) \leqslant \sum_j \Pr\{(\text{first } i \to j \text{ transition by } h) \text{ and (subsequently left } j \text{ by } h)\}$$

$$b(h) \leqslant \sum_j \int_{\sigma=0}^h v_i P_{ij} e^{-v_i \sigma}(1 - e^{-v_j(h-\sigma)})d\sigma$$

$$b(h) \leqslant \sum_j v_i P_{ij}\left[\int_{\sigma=0}^h e^{-v_i \sigma}d\sigma - \int_{\sigma=0}^h e^{-v_i \sigma}(e^{-v_j(h-\sigma)})d\sigma\right]$$

$$b(h) \leqslant \sum_j v_i P_{ij}[(h + o(h)) - (h + o(h))] = o(h).$$

Hence, $b(h) = o(h)$. Adding terms $a(h)$ and $b(h)$ yields Lemma 1.

Lemma 2. $\lim\limits_{h \to 0} \dfrac{P_{ij}(h)}{h} = v_i P_{ij}$.

The proof follows in a similar manner:

$$P_{ij}(h) = a(h) + b(h),$$

where $a(h) = \Pr\{(\text{first } i \to j \text{ transition by } h) \text{ and (subsequently stayed in } j \text{ until } h)\}$
and $b(h) = \sum\limits_k \Pr\{(\text{first } i \to k \text{ transition by } h) \text{ and (subsequently left } k \text{ but in } j \text{ at } h)\}$
where k states are adjacent to state i. Note that the summation on k includes state j.
Term $a(h)$ is one term of the summation in the first lemma and is given by

$$a(h) = \int_{\sigma=0}^{h} v_i P_{ij} e^{-v_i \sigma}(e^{-v_j(h-\sigma)}) d\sigma$$

$$= v_i P_{ij} \left\{ \frac{e^{-v_j h} - e^{-v_i h}}{v_i - v_j} \right\}$$

$$= v_i P_{ij} \left\{ \frac{1 - v_j h + o(h) - 1 + v_i h + o(h)}{v_i - v_j} \right\}$$

$$= v_i h P_{ij} + o(h).$$

Term b is bounded by

$$b(h) \leqslant \sum_k \Pr\{(\text{first } i \to k \text{ transition by } h) \text{ and (subsequently left } k)\}$$

$$b(h) \leqslant \sum_k \int_{\sigma=0}^{h} v_i P_{ik} e^{-v_i \sigma}(1 - e^{-v_k(h-\sigma)}) d\sigma$$

$$b(h) \leqslant \sum_k v_i P_{ik} \left[\int_{\sigma=0}^{h} e^{-v_i \sigma} d\sigma - \int_{\sigma=0}^{h} e^{-v_i \sigma}(e^{-v_k(h-\sigma)}) d\sigma \right]$$

$$b(h) \leqslant \sum_k v_i P_{ik}[(h + o(h)) - (h + o(h))] = o(h).$$

Hence, $b(h) = o(h)$. Adding terms $a(h)$ and $b(h)$ yields $P_{ij}(h) = v_i h P_{ij} + o(h)$. Dividing throughout by h and taking the limit yields Lemma 2.

We can now prove the CKFEs. We start by writing a two-step expression for $P_{ij}(t + h)$, where the channel goes from $i \to k$ in t, and then goes from $k \to j$ in the remaining h. Summed over all k, this must incorporate all the ways of going from $i \to j$ in $t + h$. We thus have

$$P_{ij}(t + h) = \sum_{k \in \text{ all states}} P_{ik}(t) P_{kj}(h)$$

$$= \sum_{k \in \text{ all states} \neq j} P_{ik}(t) P_{kj}(h) + P_{ij}(t) P_{jj}(h).$$

We now can write an expression for the derivative of $P_{ij}(t + h)$:

$$\frac{dP_{ij}(t)}{dt} = \lim_{h\to 0}\left\{\frac{P_{ij}(t+h) - P_{ij}(t)}{h}\right\}$$

$$\underset{\text{substituting from above}}{\equiv} \lim_{h\to 0}\left\{\sum_{k\in \text{ all states }\neq j} P_{ik}(t)\frac{P_{kj}(h)}{h} - \left[\frac{1 - P_{jj}(h)}{h}\right]P_{ij}(t)\right\}.$$

Assuming we can interchange limit with summation (which is possible for finite-state Markov models), we obtain

$$\frac{dP_{ij}(t)}{dt} = \sum_{k\in \text{ all states }\neq j} P_{ik}(t)\lim_{h\to 0}\left[\frac{P_{kj}(h)}{h}\right] - \lim_{h\to 0}\left[\frac{1 - P_{jj}(h)}{h}\right]P_{ij}(t).$$

Substituting from the two lemmas yields

$$\frac{dP_{ij}(t)}{dt} = \sum_{k\in \text{ all states }\neq j} P_{ik}(t)v_k P_{kj} - v_j P_{ij}(t).$$

Putting this into matrix form yields the CKFEs:

$$d\underbrace{\begin{bmatrix} P_{11}(t) & P_{12}(t) & \cdots & P_{1N}(t) \\ P_{21}(t) & P_{22}(t) & \cdots & P_{2N(t)} \\ \vdots & \vdots & \ddots & \vdots \\ P_{N1}(t) & P_{N2}(t) & \cdots & P_{NN}(t) \end{bmatrix}}_{\frac{d\mathbf{P}(t)}{dt}} =$$

$$\underbrace{\begin{bmatrix} P_{11}(t) & P_{12}(t) & \cdots & P_{1N}(t) \\ P_{21}(t) & P_{22}(t) & \cdots & P_{2N(t)} \\ \vdots & \vdots & \ddots & \vdots \\ P_{N1}(t) & P_{N2}(t) & \cdots & P_{NN}(t) \end{bmatrix}}_{\mathbf{P}(t)} \underbrace{\begin{bmatrix} -v_1 & v_1 P_{12} & \cdots & v_1 P_{1N} \\ v_2 P_{21} & -v_2 & & \\ \vdots & & \ddots & \\ v_N P_{N1} & v_N P_{N2} & \cdots & -v_N \end{bmatrix}}_{\mathbf{Q}}.$$

The meaning of the rate constants is clear from the elements in $\mathbf{Q}$. The relationship between the rate constants and the 'ever' probabilities is also clear. Another form of this can be derived by multiplying both sides of the above equation by a row probability vector $\mathbf{P}_a(0)$ signifying an arbitrary initial condition of the channel, where

$$\mathbf{P}_a(0) = [P_1(t=0)\; P_2(0) \ldots P_N(0)].$$

Recognizing that

$$\mathbf{P}_a(0)\mathbf{P}(t) = \mathbf{P}_a(t)$$
$$= [P_{P_a(0)1}(t)\; P_{P_a(0)2}(t)\; \cdots\; P_{P_a(0)N}(t)],$$

where $P_{P_a(0)i}(t) = \Pr\{\text{in state } i \text{ at } t \mid \text{initial condition } \mathbf{P}_a(0) \text{ at time zero}\}$, we can write an alternate form of the CKFEs:

$$\frac{d\left[\,P_{P_a(0)1}(t)\underbrace{\phantom{P_{P_a(0)2}(t)}}_{\frac{dP_a(t)}{dt}}P_{P_a(0)2}(t)\;\ldots\;P_{P_a(0)N}(t)\right.}{}$$

$$\underbrace{\left[\,P_{P_a(0)1}(t)\,P_{P_a(0)2}(t)\ldots P_{P_a(0)N}(t)\right]}_{P_a(t)}[\mathbf{Q}].$$

Oftentimes, we omit the $P_a(0)$ from the subscript, but we implicitly mean that the conditional probabilities in question are taken in the context of some arbitrary initial probability row vector.

To show that CKFEs $\Rightarrow$ exponential dwell times in a state, we argue as follows: by the definition of a lifetime, the lifetime distribution for state i is equivalent to $P_{ii}(t)$ in the CKFEs when there is no return to state i from adjacent states j. This is accomplished by setting all off-main-diagonal elements to zero in the $\mathbf{Q}$ matrix. Then $\frac{dP_{ii}(t)}{dt} = -v_i\,P_{ii}(t)[$, so that lifetimes are exponentially distributed in state i. The meaning of the traditional rate constants is clear from the form of the elements in matrix $\mathbf{Q}$.

Remark 1. *There is another form of the Chapman–Kolmogorov equation, known as the backward equation. We will not derive it here because it is not very intuitive from a mechanistic point of view. However, mathematicians like it because the exchange of summation and limit is valid even for infinite-state Markov processes.*

Example 7. *Consider the following simple gating model:*

$$C_1 \underset{k_{21}}{\overset{k_{12}}{\rightleftharpoons}} C_2 \underset{k_{32}}{\overset{k_{23}}{\rightleftharpoons}} O_3,$$

where $\mathbf{P} = [P_{C_1}(t)P_{C_2}(t)O_3(t)]$. *The matrix* $\mathbf{Q}$ *is*

$$\begin{bmatrix} -k_{12} & k_{12} & 0 \\ k_{21} & -(k_{21}+k_{23}) & k_{23} \\ 0 & k_{32} & -k_{32} \end{bmatrix}.$$

3.3.3 Solution to the Chapman–Kolmogorov forward equations

Because many of the laws of algebra and differentiation apply equally to scalar and matrix equations, we can write the solution to the CKFEs by analogy to the solution of the scalar equation $\frac{dP}{dt} = P \times (-k) \Rightarrow P(t) = \exp(-kt)$. Hence, the solution to the CFKEs is

$$\mathbf{P}(t) = [e^{\mathbf{Q}t}],$$

where $[e^{Qt}]$ is the matrix exponential and the ijth element of $[e^{Qt}]$ is $P_{ij}(t)$. What makes the scalar $\exp(-kt)$ function a solution is that term-by-term differentiation of the series definition shows that this function satisfies the differential equation. Likewise, by defining $[e^{Qt}]$ using the analogous matrix series definition, we can use term-by-term differentiation to show that it satisfies the matrix differential equation. Thus,

$$[e^{Qt}] \equiv \mathbf{I} + \mathbf{Q}t + \frac{\mathbf{Q}^2 t^2}{2!} + \ldots.$$

Similarly, the solution for $\mathbf{P}_a(t)$ is

$$\mathbf{P}_a(t) = \mathbf{P}_a(0)[e^{Qt}],$$

where the pre-exponential term is required to satisfy the initial condition at time zero.

References

Aldrich R W, Corey D P and Stevens C F 1983 A reinterpretation of mammalian sodium channel gating based on single channel recording *Nature* **306** 436–41

Colquhoun D and Hawkes A G 1981 On the stochastic properties of single ion channels *Proc. R. Soc. Lond. B Biol. Sci.* **211** 205–35

Herzig S, Patil P, Neumann J, Staschen C M and Yue D T 1993 Mechanisms of beta-adrenergic stimulation of cardiac Ca^{2+} channels revealed by discrete-time Markov analysis of slow gating *Biophys. J.* **65** 1599–612

Mason S J 1953 Feedback theory-some properties of signal flow graphs *Proc. IRE* **41** 1144–56

IOP Publishing

Ion Channel Gating and Mechanisms

David T Yue, Manu Ben-Johny and Ivy E Dick

Chapter 4

Markovian gating as a linear system II

4.1 Continuous-time Markov processes

4.1.1 A convenient property of the matrix exponential

From the properties of the exponential function, which carry over to the matrix exponential, we will show that

$$\mathbf{P}(t_2) = \mathbf{P}(t_1)[e^{Q(t_2 - t_1)}]. \tag{4.1}$$

Proof: because the matrix exponential satisfies the Chapman–Kolmogorov forward equations (CKFEs), it is certainly true that

$$\mathbf{P}(t_2) = \mathbf{P}(t = 0)[e^{Q(t_2)}].$$

Applying the properties of the matrix exponential,

$$\mathbf{P}(t_2) = \underbrace{\mathbf{P}(t = 0)[e^{Q(t_1)}]}_{\mathbf{P}(t_1)}[e^{Q(t_2 - t_1)}].$$

Making the substitution from the underscored bracket yields the relation in equation (4.1).

4.1.2 Canonical form of the Chapman–Kolmogorov forward equations

Like the discrete-step form of the CKFEs, there is a canonical form for the continuous-time CKFEs, expressed in terms of absorbing $\mathcal{A}$ and transient $\mathcal{T}$ class states. The advantage is similar: for the purposes of many calculations, we need only consider a small portion of the overall Q matrix. The canonical form of the CKFEs is:

$$\frac{d}{dt}[\mathbf{P}_\mathcal{T}(t)\vdots\mathbf{P}_\mathcal{A}(t)] = [\mathbf{P}_\mathcal{T}(t)\vdots\mathbf{P}_\mathcal{A}(t)]\begin{bmatrix} \mathbf{Q}_{TT} \vdots \mathbf{Q}_{TA} \\ \cdots \vdots \cdots \\ \mathbf{Q}_{AT} \vdots \mathbf{Q}_{AA} \end{bmatrix}.$$

doi:10.1088/978-0-7503-2388-8ch4

This form has a few notable features:

(1) A careful look at what $Q_{\mathcal{T}\mathcal{T}}$ means shows that it is sufficient to determine all the behavior of kinetics within $\mathcal{T}$ states. Specifically,

$$\frac{d}{dt}[\mathbf{P}_{\mathcal{T}}(t)] = [\mathbf{P}_{\mathcal{T}}(t)][\mathbf{Q}_{\mathcal{T}\mathcal{T}}]$$

and

$$\mathbf{P}_{\mathcal{T}}(t) = \mathbf{P}_{\mathcal{T}}(0)[e^{\mathbf{Q}_{\mathcal{T}\mathcal{T}}t}].$$

(2) According to the definition of a transient class of states, $\lim\limits_{t\to\infty} [e^{\mathbf{Q}_{\mathcal{T}\mathcal{T}}t}] = [0]$.

(3) According to the definition of an absorbing class of states, $Q_{\mathcal{A}\mathcal{T}} = [0]$.

Example 1. *Suppose we want to calculate the probability density function for the closed durations (periods between openings) for the model shown below:*

$$C_1 \underset{k_{21}}{\overset{k_{12}}{\rightleftharpoons}} C_2 \underset{k_{32}}{\overset{k_{23}}{\rightleftharpoons}} O_3.$$

We are seeking the properties of t_{close}, i.e. the combined lifetime of visits to states C_1 and C_2 without interruption by visits to O_3. t_{close} must be specified completely by parameters associated with the exits from these states and not by parameters associated with exits from other states (O_3). Otherwise, we would violate the Markovian assumption that future behavior depends only upon the state being occupied at the current time. Hence, the properties of t_{close} must be independent of k_{32}. For the convenience of calculating t_{close}, we might as well set $k_{32} = 0$ and treat O_3 as if it were an absorbing state, even though it is not an absorbing state according to a literal interpretation. In this case, $\mathcal{T} = \{C_1, C_2\}$:

$$C_1 \underset{k_{21}}{\overset{k_{12}}{\rightleftharpoons}} C_2 \overset{k_{23}}{\rightarrow} O_3. \tag{E1.1.1}$$

The advantage of reformulating the model in this way is that $P_{t_{\text{close}}}(t)$, the $\Pr\{t_{\text{close}} > t\}$, is now conveniently tabulated as

$$P_{t_{\text{close}}}(t) = 1 - P_{O_3}(t),$$

where the initial condition at $t = 0$ for the system in scheme (E1.1.1) is that the channel is in C_2 with certainty. The latter initial condition coincides with the fact that all closed periods start just after an $O_3 \to C_2$ transition. Applying the relationship between the probability density function and the complement of the distribution function, the desired probability density function of t_{close}, $p_{t_{\text{close}}}(t)$ is given by

$$p_{t_{\text{close}}}(t) = -\frac{dP_{t_{\text{close}}}(t)}{dt} = \frac{dP_{O_3}(t)}{dt}.$$

From the CKFEs as applied to the scheme (E1.1.1),

$$p_{t_{\text{close}}}(t) = \frac{dP_{O_3}(t)}{dt} = k_{23}P_{C_2}(t).$$

We must be able to obtain $P_{C_2}(t)$ from $[\mathbf{Q}_{TT}]$ because $[\mathbf{Q}_{TT}]$ completely specifies behavior in the transient states. In this case,

$$[\mathbf{P}_T(t)] = [P_{C_1}(t) P_{C_2}(t)]$$

$$[\mathbf{Q}_{TT}] = \begin{bmatrix} -k_{12} & k_{12} \\ k_{21} & -(k_{21} + k_{23}) \end{bmatrix}$$

$$[P_{C_1}(t) P_{C_2}(t)] = [0\ 1][e^{Q_{TT}t}].$$

The desired density function is then given by

$$p_{t_{\text{close}}}(t) = k_{23}P_{C_2}(t) = [0\ 1][e^{Q_{TT}t}]\begin{bmatrix} 0 \\ 1 \end{bmatrix}k_{23}$$

$$= \underbrace{[0\ 1]}_{\text{initial condition}} [e^{Q_{TT}t}] \underbrace{\begin{bmatrix} 0 \\ k_{23} \end{bmatrix}}^{Q_{TA}}, \tag{E1.2}$$

where we have expressed it in the bottom form as a prelude to the general formula presented in the paper by Colquhoun and Hawkes (1981).

4.1.3 Practical calculation of the matrix exponential using Laplace transformation

It turns out that the Laplace transformation can be applied to a matrix function by simply taking the Laplace transform of the individual elements. Hence, just as Laplace transforms greatly aid the solution of scalar differential equations, Laplace transforms also greatly aid the solution of matrix differential equations:

$$\frac{d\mathbf{P}(t)}{dt} = \mathbf{P}(t)[\mathbf{Q}]$$

$$s\mathbf{P}(s) - \mathbf{P}(t = 0) = \mathbf{P}(s)[\mathbf{Q}]$$

$$\mathbf{P}(s)[s[\mathbf{I}] - [\mathbf{Q}]] = \mathbf{P}(t = 0)$$

$$\mathbf{P}(s) = \mathbf{P}(t = 0)[s[\mathbf{I}] - [\mathbf{Q}]]^{-1}.$$

From page section 3.24 of the Unit 3 notes, it must be that

$$[e^{\mathbf{Q}(s)}] = [s[\mathbf{I}] - [\mathbf{Q}]]^{-1}.$$

Hence, to calculate the matrix exponential, we need only take the inverse Laplace transform of $[e^{\mathbf{Q}(s)}]$, working in an element-by-element fashion (see Box 4.1). In other words,

$$[e^{\mathbf{Q}(t)}] = \mathcal{L}^{-1}\{[s[\mathbf{I}] - [\mathbf{Q}]]^{-1}\}. \tag{4.2}$$

Box 4.1. Beam me up, Scotty—an analogy for Laplace transforms.

In the series Star Trek, Captain Kirk leads his heroic team on various adventures throughout the galaxy. Montgomery 'Scotty' Scott, a brilliant engineer in Kirk's team, developed an ingenious algorithm that allows him to transport people great distances. Thus, Captain Kirk would shout 'beam me up, Scotty!' and Scotty would beam him out of many challenging situations. Similarly, the Laplace domain acts as a method by which mathematically inclined individuals may rescue themselves from intractable problems and achieve higher ground by more elegant means

—David Yue

Example 2. *Here, we use the Laplace transform method to explicitly calculate the probability density function for the closed times in the previous example.*

$$[e^{Q_{\tau\tau}(t)}] = \mathcal{L}^{-1}\{[s[I] - [Q_{\tau\tau}]]^{-1}\} \tag{E2.1a}$$

$$[e^{Q_{\tau\tau}(t)}] = \mathcal{L}^{-1}\left\{\left[\begin{bmatrix} s & 0 \\ 0 & s \end{bmatrix} - \begin{bmatrix} -k_{12} & k_{12} \\ k_{21} & -(k_{21}+k_{23}) \end{bmatrix}\right]^{-1}\right\} \tag{E2.1b}$$

$$[e^{Q_{T_\tau}(t)}] = \mathcal{L}^{-1}\left\{\begin{bmatrix} s+k_{12} & -k_{12} \\ -k_{21} & s+k_{21}+k_{23} \end{bmatrix}^{-1}\right\}$$

$$[e^{Q_{\tau\tau}(t)}] = \mathcal{L}^{-1}\left\{\frac{\begin{bmatrix} s+k_{21}+k_{23} & k_{12} \\ k_{21} & s+k_{12} \end{bmatrix}}{s^2 + \underbrace{(k_{12}+k_{21}+k_{23})}_{b}s + \underbrace{(k_{12}k_{21}-k_{12}k_{21})}_{0} + \underbrace{k_{12}k_{23}}_{c}}\right\}$$

Let

$$\lambda_1, \lambda_2 = \text{negative roots of the polynomial above,}$$

so that

$$\lambda_1 = \frac{b - \sqrt{b^2 - 4c}}{2}$$

$$\lambda_2 = \frac{b + \sqrt{b^2 - 4c}}{2},$$

where

$$\lambda_1\lambda_2 = c = k_{12}k_{23}$$
$$\lambda_1 + \lambda_2 = b = k_{12} + k_{21} + k_{23}.$$

Therefore, in terms of λ_1 and λ_2, we have that

$$[e^{Q_{TT}(t)}] = \mathcal{L}^{-1}\left\{ \frac{1}{(s+\lambda_1)(s+\lambda_2)} \begin{bmatrix} s+k_{21}+k_{23} & k_{12} \\ k_{21} & s+k_{12} \end{bmatrix} \right\}$$

$$[e^{Q_{TT}(t)}] = \begin{bmatrix} \mathcal{L}^{-1}\left\{ \dfrac{s+k_{21}+k_{23}}{(s+\lambda_1)(s+\lambda_2)} \right\} & \mathcal{L}^{-1}\left\{ \dfrac{k_{12}}{(s+\lambda_1)(s+\lambda_2)} \right\} \\ \mathcal{L}^{-1}\left\{ \dfrac{k_{21}}{(s+\lambda_1)(s+\lambda_2)} \right\} & \mathcal{L}^{-1}\left\{ \dfrac{s+k_{12}}{(s+\lambda_1)(s+\lambda_2)} \right\} \end{bmatrix}.$$

Inserting this into the solution for $p_{t_{\text{close}}}(t)$ yields

$$\frac{s+k_{21}+k_{23}}{(s+\lambda_1)(s+\lambda_2)}$$

$$\frac{k_{21}}{(s+\lambda_1)(s+\lambda_2)}$$

$$\mathcal{L}^{-1}\left\{ \frac{s+k_{12}}{(s+\lambda_1)(s+\lambda_2)} \right\}$$

$$p_{t_{\text{close}}}(t) = k_{23}P_{C_2}(t) = [0\ 1] \begin{bmatrix} \mathcal{L}^{-1}\left\{ \dfrac{s+k_{21}+k_{23}}{(s+\lambda_1)(s+\lambda_2)} \right\} & \mathcal{L}^{-1}\left\{ \dfrac{k_{12}}{(s+\lambda_1)(s+\lambda_2)} \right\} \\ \mathcal{L}^{-1}\left\{ \dfrac{k_{21}}{(s+\lambda_1)(s+\lambda_2)} \right\} & \mathcal{L}^{-1}\left\{ \dfrac{s+k_{12}}{(s+\lambda_1)(s+\lambda_2)} \right\} \end{bmatrix} \begin{bmatrix} 0 \\ k_{23} \end{bmatrix}$$

$$p_{t_{\text{close}}}(t) = k_{23}P_{C_2}(t) = k_{23} \times \mathcal{L}^{-1}\left\{ \frac{s+k_{12}}{(s+\lambda_1)(s+\lambda_2)} \right\}.$$

Partial fraction expansion yields

$$p_{t_{\text{close}}}(t) = k_{23}P_{C_2}(t) = k_{23} \times \mathcal{L}^{-1}\left\{ \frac{A}{(s+\lambda_1)} + \frac{B}{(s+\lambda_2)} \right\}, \tag{E2.2.2}$$

where A and B can be obtained by the Heavyside method:

$$\left[\frac{s+k_{12}}{(s+\lambda_1)(s+\lambda_2)}(s+\lambda_1) \right]_{s=-\lambda_i} = A + \underbrace{\left[\frac{B}{(s+\lambda_2)}(s+\lambda_1) \right]_{s=-\lambda_i}}_{0}$$

$$A = \left[\frac{s+k_{12}}{(s+\lambda_2)} \right]_{s=-\lambda_1} = \frac{k_{12}-\lambda_1}{(\lambda_2-\lambda_1)}$$

$$B = \left[\frac{s+k_{12}}{(s+\lambda_1)} \right]_{s=-\lambda_2} = \frac{\lambda_2-k_{12}}{(\lambda_2-\lambda_1)}.$$

Plugging these results back into equation (E2.2.2) yields

$$p_{t_{\text{close}}}(t) = k_{23}P_{C_2}(t) = \frac{k_{23}}{(\lambda_2 - \lambda_1)} \times \mathcal{L}^{-1}\left\{\frac{k_{12} - \lambda_1}{(s + \lambda_1)} + \frac{\lambda_2 - k_{12}}{(s + \lambda_2)}\right\}.$$

Recalling that $\mathcal{L}^{-1}\{\frac{1}{s+a}\} = e^{-at}$, *we obtain*

$$p_{t_{\text{close}}}(t) = k_{23}P_{C_2}(t) = \frac{k_{23}}{(\lambda_2 - \lambda_1)}\{(k_{12} - \lambda_1)e^{-\lambda_1 t} + (\lambda_2 - k_{12})e^{-\lambda_2 t}\}.$$

4.2 Application to open–shut duration and burst kinetics (Colquhoun and Hawkes 1981)

4.2.1 Three useful tricks for channel kinetics

4.2.1.1 Trick one: calculation of the expected time for $i \to j$ transitions

Suppose a channel starts in state i at time $t = 0$. Let t_{ij} be a random variable set equal to the time required for the channel to first enter state j. Let the probability density function for t_{ij} be $f_{ij}(t)$. The first trick is a convenient way to calculate the expected time of first arrival in state j, $m_{ij} = \langle t_{ij} \rangle$. According to the definition of an expectation value,

$$m_{ij} = \int_{t=0}^{\infty} t f_{ij}(t)dt = -\frac{d}{ds}\left[\int_{t=0}^{\infty} e^{-st} f_{ij}(t)dt\right]_{s=0}$$

$$m_{ij} = -\frac{d}{ds}[f_{ij}(s)]_{s=0}. \tag{4.3}$$

4.2.1.2 Trick two: canonical form for calculating $f_{ij}(t); i \in \mathcal{T}, j \in \mathcal{A}$

Suppose a channel starts in state $i \in \mathcal{T}$ at time $t = 0$. Let t_{ij} be a random variable set equal to the time required for the channel to first enter the absorbing class of states in state $j \in \mathcal{A}$. Then, generalization of equation (E1.2.2) from example 1 yields

$$[f_{ij}] = \underbrace{[e^{Q_{\mathcal{T}\mathcal{T}}(t)}]}_{i,\,k\ \text{element}=\Pr\{k\in\mathcal{T}@t|i\in\mathcal{T}@0\}}\ \overbrace{[\mathbf{Q}_{\mathcal{T}\mathcal{A}}]}^{\text{rate cons for } k\to j\in\mathcal{A}}. \tag{4.4}$$

4.2.1.3 Trick three: probability density function for the sum of independent random variables

Suppose we are interested in the probability density function for a duration T that is defined as the sum of several independent types of dwell times $t_1 \ldots t_n$. If we know the probability density function for each of the individual dwell times $f_1(t) \ldots f_n(t)$, then we can calculate the probability density function for T, which we will call $f_{\text{TOT}}(t)$. In mathematical terms,

$$T = \sum_{i=1}^{n} t_i.$$

From reasoning of the sort presented in the lecture notes for chapter 2 (section 2.6.4), we have the standard probability result that

$$f_{\text{TOT}}(t) = f_1(t) \times f_2(t) \times \ldots \times f_n(t),$$

where $\times$ is the convolution operator. One of the beautiful properties of Laplace transforms is that convolution in the time domain is equivalent to multiplication in the s domain. Hence, we have the simple relation that

$$f_{\text{TOT}}(t) = \mathcal{L}^{-1}\{f_{\text{TOT}}(s)\}$$

$$= \mathcal{L}^{-1}\{f_1(s) \times f_2(s) \times \ldots \times f_n(s)\}. \tag{4.5}$$

4.2.2 Properties of the Castillo and Katz model (Del Castillo and Katz 1957) of a ligand-gated channel

Consider the following model for a ligand-gated channel:

$$\underset{3}{T} \underset{k_{-1}}{\overset{k_1}{\rightleftharpoons}} \underset{2}{AT} \underset{\alpha}{\overset{\beta}{\rightleftharpoons}} \underset{1}{AR},$$

where $k_1 = \overline{k_1}$ (agonist), and AR is the only conducting state. β and α are often much faster than k_1 and k_{-1}. This tends to make the channel burst, as shown in figure 4.1.

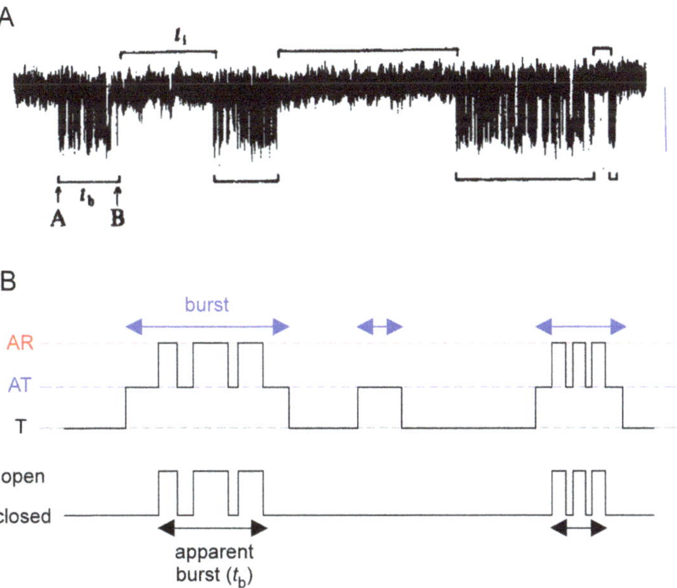

Figure 4.1. Bursting behavior of a ligand-gated ion channel patch. Despite the presence of multiple channels, one-level bursts are likely attributable to the gating of a single channel. (A) Data illustrating bursting in a single channel patch. Data is recorded from an acetylcholine-activated channel in frog muscle. Reproduced from [Sakmann *et al* 1980], with permission from Springer Nature. (B) The burst structure reflects rapid transitions between the AT and AR states, with AR being the only conducting state.

Often, it is difficult to get just one channel in the patch. It is therefore difficult to interpret periods between bursts because these are greatly influenced by the number of channels in the patch. However, a one-level burst is almost certainly due to the gating of one of the channels. Hence, if we could glean information about gating simply from the kinetics of one-level bursts, we could still garner useful information from this multichannel patch. This sort of analysis is termed 'burst analysis.' The diagram in figure 4.1 shows the structure of typical bursts.

4.2.2.1 Number of openings per burst

We wish to calculate the $\Pr\{k$ openings per burst$\} = P(k)$, where a burst starts in state 2. We consider the following signal-flow diagram.

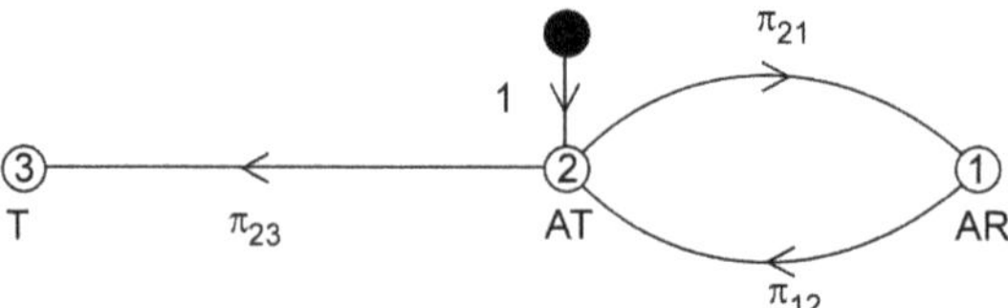

From simple inspection, $P(k) = (\pi_{21}\pi_{12})^k \pi_{23}$, where $\pi_{21} = \beta/(\beta + k_{-1})$, $\pi_{12} = 1$, and $\pi_{23} = k_{-1}/(\beta + k_{-1})$ (from the CKFE derivation). The mean number of openings per burst $\langle k \rangle$ can be solved the standard way:

$$\langle k \rangle = \sum_{j=0}^{\infty} jP(j) = \pi_{23} \sum_{j=0}^{\infty} j\pi_{21}^j = \pi_{23} \frac{\pi_{21}}{(1 - \pi_{21})^2}$$

$$= \pi_{23} \frac{\pi_{21}}{\pi_{23}^2} = \frac{\pi_{21}}{\pi_{23}} = \frac{\beta}{k_{-1}}.$$

For the mean number of openings in an apparent burst, we must consider a different starting point.

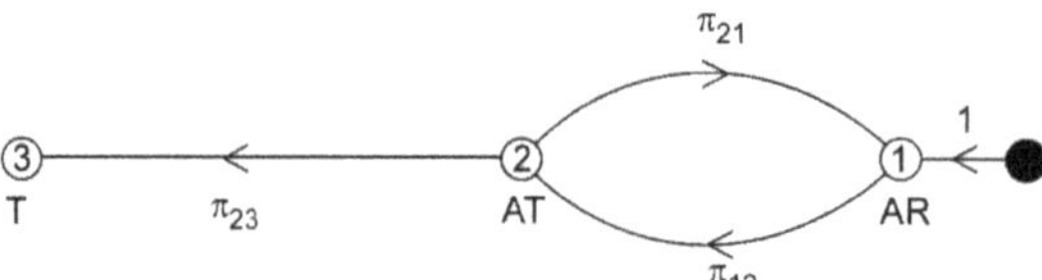

From simple inspection, $P_{\text{app}}(k) = \Pr\{k$ openings $|{\geqslant}1$ opening$\} = \pi_{12}^k \pi_{21}^{k-1} \pi_{23} = \pi_{21}^{k-1} \pi_{23}$. Another way of calculating this, using the basic definitions of conditional probability, is:

$$P_{\text{app}}(k) = \Pr\{k \text{ openings} |{\geqslant}1 \text{ opening}\}$$

$$= \frac{\Pr\{(k \text{ openings}) \text{ and } ({\geqslant}1 \text{ opening})\}}{\Pr\{{\geqslant}1 \text{ opening}\}}$$

$$= \frac{(\pi_{21}\pi_{12})^k \pi_{23}}{\pi_{21}} = \pi_{12}^k \pi_{21}^{k-1} \pi_{23} = \pi_{21}^{k-1} \pi_{23} \text{ for } k \geqslant 1.$$

The mean number of openings per apparent burst $\langle k \rangle_{\mathrm{app}}$ can be solved in the standard way:

$$\langle k \rangle_{\mathrm{app}} = \sum_{j=0}^{\infty} j P_{\mathrm{app}}(j) = \pi_{23} \sum_{j=0}^{\infty} j \pi_{21}^{j-1} = \pi_{23} \frac{1}{(1 - \pi_{21})^2}$$

$$= \pi_{23} \frac{1}{\pi_{23}^2} = \frac{1}{\pi_{23}} = \frac{\beta}{k_{-1}} + 1 = \langle k \rangle + 1.$$

Alternatively, we could get both actual and apparent expectations for the number of openings using the fundamental matrix:

$$\mathbf{P}_T(k) = [P_1(k)P_2(k)]$$

$$\boldsymbol{\pi}_{TT} = \begin{bmatrix} 0 & 1 \\ \dfrac{\beta}{(\beta + k_{-1})} & 0 \end{bmatrix}$$

$$M = [\mathbf{I} - \boldsymbol{\pi}_{TT}]^{-1}$$

$$= \begin{bmatrix} 1 & -1 \\ -\dfrac{\beta}{(\beta + k_{-1})} & 1 \end{bmatrix}^{-1}$$

$$= \frac{\beta + k_{-1}}{k_{-1}} \begin{bmatrix} 1 & 1 \\ \dfrac{\beta}{(\beta + k_{-1})} & 1 \end{bmatrix}.$$

The 1,1 element gives $\langle k \rangle_{\mathrm{app}}$, and the 2,1 element gives $\langle k \rangle$.

4.2.2.2 Probability density function of burst length

Let us set up the continuous-time system with $\mathbf{P}(t) = [P_1(t)P_2(t)P_3(t)]$. Hence,

$$\mathbf{Q} = \begin{bmatrix} -\alpha & \alpha & \vdots & 0 \\ \beta & -(k_{-1} + \beta) & \vdots & k_{-1} \\ \cdots & \cdots & \vdots & \cdots \\ 0 & (k_1) & \vdots & (-k_1) \end{bmatrix},$$

and if in our mind's eye, we make state 3 absorbing by setting the parenthetical elements in the bottom row of partitions in $\mathbf{Q}$ to zero, we have

$$\mathbf{Q}_{TT} = \begin{bmatrix} -\alpha & \alpha \\ \beta & -(k_{-1} + \beta) \end{bmatrix}.$$

From trick 2 in section 4.2.1, we have that the desired entities are, in principle,

$$\begin{bmatrix} f_{13}(t) \\ f_{23}(t) \end{bmatrix} = [e^{\mathbf{Q}_{TT}(t)}] \underbrace{\begin{bmatrix} 0 \\ k_{-1} \end{bmatrix}}_{\mathbf{Q}_{TA}}.$$

Now, let us calculate $[e^{\mathbf{Q}_{TT}(t)}]$:

$$[e^{\mathbf{Q}_{TT}(s)}] = [s\mathbf{I} - \mathbf{Q}_{TT}]^{-1}$$

$$= \begin{bmatrix} s + \alpha & -\alpha \\ -\beta & s + k_{-1} + \beta \end{bmatrix}^{-1}$$

$$= \frac{\begin{bmatrix} s + k_{-1} + \beta & \alpha \\ \beta & s + \alpha \end{bmatrix}}{s^2 + \underbrace{(\alpha + k_{-1} + \beta)}_{b} s + \underbrace{\alpha k_{-1}}_{c}}.$$

Let λ_1, $\lambda_2 = $ the negative roots of the polynomial above, such that

$$\lambda_1 = \frac{b - \sqrt{b^2 - 4c}}{2}$$

$$\lambda_2 = \frac{b + \sqrt{b^2 - 4c}}{2},$$

where

$$\lambda_1 \lambda_2 = c = \alpha k_{-1}$$
$$\lambda_1 + \lambda_2 = b = \alpha + k_{-1} + \beta.$$

So, in terms of λ_1 and λ_2, we have that

$$[e^{\mathbf{Q}_{TT}(t)}] = \mathcal{L}^{-1}\left\{ \frac{1}{(s + \lambda_1)(s + \lambda_2)} \begin{bmatrix} s + k_{-1} + \beta & \alpha \\ \beta & s + \alpha \end{bmatrix} \right\}$$

$$[e^{\mathbf{Q}_{TT}(t)}] = \begin{bmatrix} \mathcal{L}^{-1}\left\{ \dfrac{s + k_{-1} + \beta}{(s + \lambda_1)(s + \lambda_2)} \right\} & \mathcal{L}^{-1}\left\{ \dfrac{\alpha}{(s + \lambda_1)(s + \lambda_2)} \right\} \\ \mathcal{L}^{-1}\left\{ \dfrac{\beta}{(s + \lambda_1)(s + \lambda_2)} \right\} & \mathcal{L}^{-1}\left\{ \dfrac{s + \alpha}{(s + \lambda_1)(s + \lambda_2)} \right\} \end{bmatrix}.$$

Plugging this into the expression for $\begin{bmatrix} f_{13}(t) \\ f_{23}(t) \end{bmatrix}$, we get

$$\begin{bmatrix} f_{13}(t) \\ f_{23}(t) \end{bmatrix} = \begin{bmatrix} \mathcal{L}^{-1}\left\{ \dfrac{s + k_{-1} + \beta}{(s + \lambda_1)(s + \lambda_2)} \right\} & \mathcal{L}^{-1}\left\{ \dfrac{\alpha}{(s + \lambda_1)(s + \lambda_2)} \right\} \\ \mathcal{L}^{-1}\left\{ \dfrac{\beta}{(s + \lambda_1)(s + \lambda_2)} \right\} & \mathcal{L}^{-1}\left\{ \dfrac{s + \alpha}{(s + \lambda_1)(s + \lambda_2)} \right\} \end{bmatrix} \begin{bmatrix} 0 \\ k_{-1} \end{bmatrix}$$

$$= \begin{bmatrix} k_{-1}\mathcal{L}^{-1}\left\{ \dfrac{\alpha}{(s + \lambda_1)(s + \lambda_2)} \right\} \\ k_{-1}\mathcal{L}^{-1}\left\{ \dfrac{s + \alpha}{(s + \lambda_1)(s + \lambda_2)} \right\} \end{bmatrix}$$

$$= \begin{bmatrix} \dfrac{\alpha k_{-1}(e^{-\lambda_1 t} - e^{-\lambda_2 t})}{\lambda_2 - \lambda_1} \\ \dfrac{k_{-1}}{\lambda_2 - \lambda_1}((\alpha - \lambda_1)e^{-\lambda_1 t} - (\lambda_2 - \alpha)e^{-\lambda_2 t}) \end{bmatrix}.$$

Here, $f_{23}(t)$ gives the density function for the genuine burst duration, and $f_{13}(t)$ is closely related to (but not quite equal to) the apparent burst duration.

4.2.2.3 Mean burst length

Let t_{burst} be a random variable that represents the length of a burst. What is the mean burst length?

$$\langle t_{\text{bursit}} \rangle = \int_{t=0}^{\infty} t f_{23}(t)\,dt$$

Use trick 1 in 4.2.1:

$$<t_{\text{burst}}> = -\frac{d}{ds}[f_{23}(s)]_{s=0}$$

$$= -\frac{d}{ds}\left[\frac{k_{-1}}{\lambda_2 - \lambda_1}\left(\frac{(\alpha - \lambda_1)}{(s + \lambda_1)} + \frac{(\lambda_2 - \alpha)}{(s + \lambda_2)} \right) \right]_{s=0}$$

$$= \frac{\alpha + \beta}{k_{-1}\alpha}.$$

4.2.2.4 Probability density function of apparent burst length

The duration of an apparent burst is t_{AB}, which is given by:

$$t_{AB} + t_{AT \to T} = t_{13},$$

where t_{13} is given by the $f_{13}(t)$ function above, and $t_{AT \to T}$ is the random variable for the length of a single sojourn in AT that terminates in T. What we desire is the probability density function for t_{AB}, $f_{AB}(t)$. We already know $f_{13}(t)$, as calculated above. We can calculate $f_{AT \to T}(t)$ for $t_{AT \to T}$ as follows: $f_{AT \to T}(t) = -d\Pr\{t_{AT} > t \mid AT \to T\}/dt$.

$$\Pr\{t_{AT} > t \mid AT \to T\} = \frac{\Pr\{t_{AT} > t\}\Pr\{AT \to T\}}{\Pr\{AT \to T\}}$$
$$= \Pr\{t_{AT} > t\}$$
$$= e^{-(k_{-1} + \beta)t}$$

Taking the derivative yields

$$f_{AT \to T}(t) = (k_{-1} + \beta)e^{-(k_{-1} + \beta)t}.$$

Note that $f_{AT \to T}(t) = f_{AT}(t)$. This counterintuitive finding is an instance of the 'Chinese banquet' hypothesis.

We can now calculate $f_{AB}(t)$. From equation (4.5) (trick 3 in 4.2.1), we have that

$$f_{13}(t) = f_{AB}(t) \times f_{AT \to T}(t).$$

Rather than deconvolving this to obtain $f_{AB}(t)$, let us proceed in the s domain:

$$f_{13}(s) = f_{AB}(s) \times f_{AT \to T}(s)$$

$$\begin{aligned}
f_{AB}(t) &= \mathcal{L}^{-1} f_{AB}(s) \\[4pt]
&= \mathcal{L}^{-1}\left[\frac{f_{13}(s)}{f_{AT \to T}(s)} \right] \\[4pt]
&= \mathcal{L}^{-1}\left[\frac{\frac{\alpha k_{-1}}{(s + \lambda_1)(s + \lambda_2)}}{\frac{(k_{-1} + \beta)}{(s + k_{-1} + \beta)}} \right] \\[4pt]
&= \frac{\alpha k_{-1}}{\lambda_2 - \lambda_1}\left\{ \frac{(k_{-1} + \beta - \lambda_1)}{k_{-1} + \beta}e^{-\lambda_1 t} - \frac{(k_{-1} + \beta - \lambda_2)}{k_{-1} + \beta}e^{-\lambda_2 t} \right\}.
\end{aligned}$$

4.2.2.5 Mean burst length

t_{AB} is the random variable representing the length of an apparent burst. What is the mean apparent burst length?

$$\langle t_{AB} \rangle = \int_{t=0}^{\infty} t f_{AB}(t)\,dt$$

Use trick 1 in 4.2.1:

$$\begin{aligned}
\langle t_{AB} \rangle &= -\frac{d}{ds}[f_{AB}(s)]_{s=0} \\[4pt]
&= -\frac{d}{ds}\left[\frac{\alpha k_{-1}}{(s + \lambda_1)(s + \lambda_2)} \frac{(s + k_{-1} + \beta)}{(k_{-1} + \beta)} \right]_{s=0} \\[4pt]
&= \underbrace{\frac{(\alpha + \beta + k_{-1})}{(k_{-1}\alpha)}}_{\langle t_{13} \rangle} - \underbrace{\frac{1}{(k_{-1} + \beta)}}_{\langle t_{AT \to T} \rangle}.
\end{aligned}$$

4.2.3 Mean open and shut times within a burst

In the following calculations, we are going to take advantage of the following probability relation. Let X and Y be independent random variables. It turns out that

$$\langle XY \rangle = \langle X \rangle \langle Y \rangle.$$

Proof: let $f_X(X)$ and $f_Y(Y)$ be the probability density functions for X and Y. Then,

$$\langle X \rangle = \int_{X=0}^{\infty} X f_X(X) dX$$
$$\langle Y \rangle = \int_{Y=0}^{\infty} Y f_Y(Y) dY$$
$$\langle XY \rangle = \int_{X=0}^{\infty} \int_{Y=0}^{\infty} XY f_{XY}(XY) dX dY.$$

If X and Y are independent, then

$$\langle XY \rangle = \int_{X=0}^{\infty} \int_{Y=0}^{\infty} XY f_X(X) f_Y(Y) dX dY$$
$$= \left(\int_{X=0}^{\infty} X f_X(X) dX \right) \left(\int_{Y=0}^{\infty} Y f_Y(Y) dY \right)$$
$$= \langle X \rangle \langle Y \rangle.$$

4.2.3.1 Mean total open time per burst
Let k be the number of openings per burst. Let t_{AR} be the length of openings in a burst. Because the two random variables are independent, the mean total open time per burst is given by

$$\langle k t_{AR} \rangle = \langle k \rangle \langle t_{AR} \rangle$$
$$= \frac{\beta}{k_{-1}} \frac{1}{\alpha}.$$

In this example and the following, the dwell time in any state and the number of overall visits to that state must be independent, in accordance with the Markovian assumption. Otherwise, the dwell time would depend on the number of visits in the past or the future.

4.2.3.2 Mean total shut time per burst
Let k_{shut} be the number of shut periods (number of sojourns in AT) per burst. Let t_{AT} be the length of sojourns in AT in a burst. We are interested in the average value of $k_{\text{shut}} \times t_{AT}$. Note that $k_{\text{shut}} = k + 1$.

$$\langle (k + 1) t_{AT} \rangle$$

Because the two random variables are independent,

$$<(k+1)t_{AT}> = \langle k+1 \rangle <t_{AT}>$$
$$= (\langle k \rangle + 1)<t_{AT}>$$
$$= \left(\frac{\beta}{k_{-1}} + 1\right)\frac{1}{(\beta + k_{-1})}$$
$$= \frac{1}{k_{-1}}.$$

This is remarkable; it is as if time stops during sojourns in AR.

4.2.3.3 *Mean total open time per apparent burst*

Let k_{app} be the apparent number of openings per burst. We want to find $\langle k_{\text{app}} t_{AR} \rangle$. The two random variables are independent, so we have

$$<k_{\text{app}} t_{AR}> = \langle k_{\text{app}} \rangle <t_{AR}>$$
$$= \left(\frac{\beta}{k_{-1}} + 1\right)\frac{1}{\alpha}.$$

4.2.3.4 *Mean total shut time per apparent burst*

Let $k_{\text{shut, app}}$ be the number of shut periods (sojourns in AT between visitations to AR) per burst. We are interested in $\langle k_{\text{shut, app}} t_{AT} \rangle$. Because the two random variables are independent, we have

$$<k_{\text{shut, app}} t_{AT}> = \langle k_{\text{shut, app}} \rangle <t_{AT}>$$
$$= \langle (k_{\text{app}} - 1) \rangle \frac{1}{(\beta + k_{-1})}$$
$$= (<k_{\text{app}}> - 1)\frac{1}{(\beta + k_{-1})}$$
$$= \frac{\beta}{k_{-1}}\frac{1}{(\beta + k_{-1})}.$$

4.2.4 How can one tell whether a shut period falls within a burst?

The basic idea is to calculate the probability that if a shut period has length $<T$, what is the probability that it is within a burst? If we choose T such that it is appropriately short, then most of the shut periods lasting for $\leqslant T$ fall within a burst. Therefore, we want to calculate the mysterious entity $\text{Pr}\{\text{shut period is within a burst} \mid t_{\text{shut}} < T\}$

$$\text{Pr}\left\{\underbrace{\text{shut period is within a burst}}_{A} \mid \underbrace{t_{\text{shut}} < T}_{B}\right\}.$$

Taking advantage of Bayes's formula,

$$\Pr\{A \mid B\} = \frac{\Pr\{B \mid A\}\Pr\{A\}}{\Pr\{B\}}.$$

Applying this to our situation yields $\Pr\{\underbrace{\text{shut period is within a burst}}_{A} \mid \underbrace{t_{\text{shut}} < T}_{B}\}$

$$= \frac{\Pr\{\underbrace{t_{\text{shut}} < T}_{B} \mid \underbrace{\text{shut period is within a burst}}_{A}\}\Pr\{\underbrace{\text{shut period is within a burst}}_{A}\}}{\Pr\{\underbrace{t_{\text{shut}} < T}_{B}\}}$$

$$= \frac{\{1 - F_{AT}(T)\}\{\pi_{AT \to AR}\}}{\{1 - F_{21}(T)\}}$$

$$= \frac{e^{-(\beta + k_{-1})T}(\beta/(\beta + k_{-1}))}{\left(1 - \int_{\sigma=0}^{T} f_{21}(\sigma)d\sigma\right)},$$

where f_{21} is a biexponential function.

4.2.5 Constraints on gating parameters of channels that do not consume energy

Most ionic channels are not expected to consume energy in the course of their gating. If channels are allowed to 'get used to' a given voltage or ligand concentration, they reach a steady-state condition (when $P_i(t)$ no longer changes with increasing t), which is also a thermodynamic equilibrium condition. If, for example, gating transitions are stoichiometrically linked to adenosine triphosphate (ATP) hydrolysis (as in the cystic fibrosis transmembrane conductance regulator (CFTR) channel), then the steady state is not tantamount to equilibrium. Garden-variety channels are probably at equilibrium. For these channels, there are very powerful constraints and tools for gating, which supplement the kinetic analysis that we have discussed thus far.

4.2.5.1 Definition of the Boltzmann distribution

At thermodynamic equilibrium, the states of a system (e.g. a channel) have an equilibrium probability given by the Boltzmann distribution, where, for an arbitrary state i,

$$P_i(\infty) = \frac{e^{-G_i/RT}}{\sum\limits_{j \in \text{ all states}} e^{-G_j/RT}} \underset{\text{can always define G relative to arbitrary reference state 1}}{=\!=\!=} \frac{e^{-G_{1i}/RT}}{\sum\limits_{j \in \text{ all states}} e^{-G_{1j}/RT}},$$

where G_{1i} is the free-energy change when a mole of channels transitions from state 1 to state i.

4.2.5.2 *Boltzmann distribution requires microscopic reversibility condition for channels at equilibrium*

The Boltzmann distribution imposes important constraints on the gating structure of channels that do not consume energy in the course of gating. The first constraint concerns the ratio of forward and backward rate constants between any two arbitrary adjacent states. Consider the pair of states:

$$S_1 \underset{k_{21}}{\overset{k_{12}}{\rightleftharpoons}} S_2.$$

At equilibrium, the ratio of the equilibrium probabilities for these two states is given by the following simple expression as a result of the Boltzmann distribution:

$$\frac{P_{S_2}(\infty)}{P_{S_1}(\infty)} = \frac{\dfrac{e^{-G_2/RT}}{\displaystyle\sum_{j \in \text{ all states}}}}{\dfrac{e^{-G_1/RT}}{\displaystyle\sum_{j \in \text{ all states}} e^{-G_j/RT}}} = e^{-(G_2 - G_1)/RT} = e^{-\Delta G_{12}/RT}.$$

Note that from consideration of the CKFEs,

$$\frac{P_{S_2}(\infty)}{P_{S_1}(\infty)} = \frac{k_{12}}{k_{21}}.$$

Putting these two relations together, we get the following constraint on rate constants between two arbitrary states:

$$\frac{P_{S_2}(\infty)}{P_{S_1}(\infty)} = \frac{k_{12}}{k_{21}} = e^{-\Delta G_{12}/RT}.$$

Now consider two states i and j separated by an arbitrary number of transitions:

$$S_i \rightleftharpoons S_x \rightleftharpoons \cdots \rightleftharpoons S_y \rightleftharpoons S_j.$$

The second constraint arises from the fact that the Boltzmann distribution makes no distinction between adjacent or distant states, so that the ratio of equilibrium probabilities for states i and j is still given by

$$\frac{P_{S_j}(\infty)}{P_{S_i}(\infty)} = e^{-\Delta G_{ij}/RT}.$$

Consideration of the CKFEs also requires that the ratio of the product of forward and backward rate constants be constrained by

$$\frac{P_{S_j}(\infty)}{P_{S_i}(\infty)} = \frac{k_{ix}}{k_{xi}} \cdots \frac{k_{yj}}{k_{jy}}.$$

Hence, the second constraint,

$$\frac{P_{S_j}(\infty)}{P_{S_i}(\infty)} = \frac{k_{ix}}{k_{xi}} \cdots \frac{k_{yj}}{k_{jy}} = e^{-\Delta G_{ij}/RT}.$$

The latter condition, which does not require that $i \neq j$, imposes an extremely important constraint on the rate constants for any loop in a gating diagram. When setting $i = j$, we get the third and most powerful constraint.

4.2.5.3 Microscopic reversibility

For channels that do not consume energy during gating, the ratio of the products of forward and backward rate constants around any loop must be equal to unity. This constraint arises because $P_{S_i}(\infty)/P_{S_i}(\infty) = 1$ by definition or because $\Delta G_{ii} = 0$ by definition.

Is there any practical benefit to the Boltzmann perspective? It turns out that using the perspective of the Boltzmann distribution to predict the behavior of a channel at equilibrium often is far easier than fighting it out from the rate-constant perspective; the mathematics just works out in a more natural manner for questions concerning equilibrium.

4.2.5.4 Origin of the Boltzmann distribution

As Donald McQuarrie puts it, 'the Boltzmann Distribution is one of the most fundamental and useful quantities in physical chemistry.' (McQuarrie, 1975) Where does this magical formula come from? It comes from three fundamental constraints.

To understand the first constraint, we have to consider the number of ways of distributing an ensemble of channels (each distinguishable) among M states, where a channel in state i contributes a free energy E_i to the ensemble. For concreteness, let us take a specific example in which there are ten states, composed of five groups of two states, in which the E_s for the states in a group are identical. Suppose there are four channels in the ensemble. Then, the numbers of ways of populating the states with group occupancy numbers $(N_1 \ldots N_5)$ are given in the table below.

Distribution (N_1, N_2, N_3, N_4, N_5)	Number of states t (Boltzmann)
(1, 0, 2, 0, 1)	192
(1, 1, 0, 1, 1)	384
(0, 1, 2, 1, 0)	192
(0, 0, 4, 0, 0)	16
(0, 2, 0, 2, 0)	96
(1, 0, 1, 2, 0)	192
...	...

The numbers can be calculated from the formula below, where t is the number of ways of populating states according to a given set of $\{N_1, \ldots, N_5\}$:

$$t = \frac{N!}{\prod_i N_i!} \Pi_i g_i^{N_i},$$

where N is the number of channels, N_i is the number of channels in state group i, and g_i is the number of states in group i (two in the particular example above for all i). The formula can be justified as follows: $N!/\Pi_i N_i!$ gives the number of ways that N channels can be distributed among group occupancies given by a particular set of $\{N_1, \dots, N_5\}$. Given a particular instance of distribution among groups satisfying a particular $\{N_1, \dots, N_5\}$, there are $g_i^{N_i}$ ways in which N_i channels can be distributed among the states in group i. Given a particular instance of distribution among groups satisfying a particular $\{N_1, \dots, N_5\}$, there are then $\Pi_i g_i^{N_i}$ ways in which channels can populate the states within all occupied groups. We are now ready to state the first constraint:

Constraint 1: According to the core assumption of statistical thermodynamics, at thermodynamic equilibrium, the equilibrium probability of state i is given by $P_{k \in \text{group } i}(\infty) = (N_i/g_i)/N$ for the set of group occupancies $\{N_1, \dots, N_5\}$ with the largest t. The assumption is essentially radical: channels can populate states randomly, with equal likelihood. Out of this blind voting strategy (from an individual channel's perspective), some population profiles can occur in more ways than others. Equilibrium is postulated to occur when the system approximates the population profile $\{N_1, \dots, N_5\}$ with the most ways of occurring.

The second and third constraints come from how we define an 'isolated' ensemble at equilibrium. The number of channels in the ensemble is usually fixed at N, and the total energy of the ensemble remains fixed at E. These give rise to the following:

Constraint 2: $N = \sum_i N_i$

Constraint 3: $E = \sum_i N_i E_i.$

If we maximize t, subject to constraints 2–3, we obtain the Boltzmann distribution above. The method for doing this is called Lagrange multipliers.

References

Colquhoun D and Hawkes A G 1981 On the stochastic properties of single ion channels *Proc. R. Soc. Lond. B Biol. Sci.* **211** 205–35

Del Castillo J and Katz B 1957 Interaction at end-plate receptors between different choline derivatives *Proc. R. Soc. Lond. B Biol. Sci.* **146** 369–81

McQuarrie D A 1975 *Statistical Mechanics* ch 4 (Dulles, VA: University Science Books)

Sakmann B, Patlak J and Neher E 1980 Single acetylcholine-activated channels show burst-kinetics in presence of desensitizing concentrations of agonist *Nature* **286** 71–3

IOP Publishing

Ion Channel Gating and Mechanisms

David T Yue, Manu Ben-Johny and Ivy E Dick

Chapter 5

Fluctuation analysis of ion-channel gating

5.1 Introduction

As noted in chapter 2, macroscopic currents measured from whole cells or from large membrane patches approximate the collective, or 'average,' behavior of the ion channels present. However, these recordings do not directly reveal the properties of individual channels. Fluctuation (noise) analysis of macroscopic currents provides a strategy that allows us to infer single-channel kinetics from ensemble data.

The variability observed in macroscopic current recordings arises from two main sources:

1. **Instrument noise**—electrical noise inherent to the recording apparatus.
2. **Stochastic channel gating**—random fluctuations that reflect the probabilistic opening and closing of individual ion channels.

By carefully delineating these contributions, one can estimate single-channel properties. This approach is particularly useful for ion channels with very small unitary conductance (where direct single-channel analysis, see chapters 3 and 4, is impractical) or for channels with gating kinetics too rapid to resolve directly. Fluctuation analysis thus provides a powerful means of probing the elementary properties of ion channels.

5.2 Basic tools for fluctuation analysis

5.2.1 Stochastic processes defined

To derive the key relationships underlying fluctuation analysis, a few probabilistic concepts are essential:

A **random variable** is a mathematical function whose domain is the sample space of all possible outcomes and whose range is a measurable set. For example, the state of an ion channel at any given time t can be modeled as a random variable with possible values {open, closed, inactive, etc.}.

doi:10.1088/978-0-7503-2388-8ch5

A **stochastic process** is a collection of random variables indexed by time, representing the time evolution of the system. For ion channels, this might describe the sequence of open/closed states over time.

A **sample function** is a single outcome or realization of the stochastic process— for instance, one particular record of channel openings and closings.

An **ensemble** is a collection of all possible sample functions of the stochastic process, i.e., a collection of recordings for a channel.

We may further categorize stochastic processes as being either **stationary** or **nonstationary**. A stationary process is one whose statistical properties, such as mean or variance, do not change over time. For example, if we consider a population of identical ion channels in a cell, stationarity implies that the probability distribution of states (e.g. the fraction of channels open) remains constant over time. In this case, any specific ion channel may still switch between different states.

5.2.2 Time-domain methods

We now introduce several key time-domain tools for characterizing stochastic processes. These apply to both stationary and nonstationary processes. Let $x(t)$ denote a random variable defined at time t:

The **mean** value or expected value of x at each time point is defined as:

$$\mu(t) = \text{mean} = \langle x(t) \rangle_i = \int_{-\infty}^{\infty} x f(x(t_1)) dx.$$

Here, f denotes the probability density function for $x(t_1)$.

Autocovariance describes the covariance of the process with itself at time points t_1 and t_2. Conceptually, this metric provides a measure of the relationship between the fluctuations at different times, i.e. the relationship with its own past values:

$$C(t_1, t_2) \equiv \text{autocovariance} \equiv <(x(t_1) - \mu(t_1))(x(t_2) - \mu(t_1))$$
$$= \iint_{-\infty}^{\infty} (x(t_1) - \mu(t_1))(x(t_2) - \mu(t_2)) \times f(x(t_1) \text{ and } x(t_2)) \times dx(t_1)dx(t_2)$$
$$= \int_{-\infty}^{\infty} x(t_1)x(t_2)f(x(t_1) \text{ and } x(t_2))dx(t_1)dx(t_2) - \mu(t_1)\mu(t_2)$$
$$= R(t_1, t_2) - \mu(t_1)\mu(t_2),$$

where $R(t_1, t_2)$ denotes the **autocorrelation**.

Variance is the autocovariance determined at the same time point. That is:

$$\sigma^2_{x(t_1)} \equiv <(x(t_1) - \mu(t_1))^2>_i \equiv C(t_1, t_1).$$

5.3 Relation between single-channel gating and time-domain tools

Consider a general model of an ion channel that can occupy one of M possible states. Each state k is characterized by a single-channel current denoted by i_k and a time-dependent probability of occupancy denoted by $p_k(t)$ with $\sum_{k=1}^{M} p_k(t) = 1$.

For a population of N_C independent and identical channels, the mean macroscopic current is the sum over all possible states:

$$\mu(t) = N_C \sum_{k=1}^{M} i_k p_k(t).$$

This expression shows that the mean current is simply the expected value of the single-channel current scaled by the number of channels in the ensemble.

The autocovariance of the macroscopic current at two times, t_1 and t_2, can be written as:

$$C(t_1, t_2) = N_c \sum_k \sum_j p_j(t_1) p_{jk}(t_2 \mid t_1)(i_j - \mu(t_1))(i_k - \mu(t_2)),$$

where $p_j(t_1)$ is the probability that the channel is in state j at time t_1, and $p_{jk}(t_2|t_1)$ is the conditional probability that the channel is in state k at time t_2, given that it was in state j at t_1. This expression can be simplified to:

$$= N_c \left[\sum_k \sum_j p_j(t_1) p_{jk}(t_2 \mid t_1)\{i_j i_k\} - \mu(t_2)\mu(t_1) \right].$$

This expression quantifies how fluctuations in the current at time t_1 are correlated with those at time t_2. The first term represents the expected product of single-channel currents, weighted by the probability of occupying state j at t_1 and having transitioned to state k by t_2. The subtraction of $\mu(t_1) \cdot \mu(t_2)$ removes the contribution from the mean currents, leaving only the correlated fluctuations.

In other words, $C(t_1, t_2)$ captures the memory of the channel's stochastic gating: how likely it is that being in one state at t_1 influences the state (and therefore the current amplitude) at a later time t_2.

Example 1. *As a concrete example, consider an ion channel with two conductance states, i.e. if state 1 represents the open state with a single-channel current amplitude $i_1 = i$ and all other states $j > 0$ are nonconductive with $i_j = 0$.*

The mean value is given by: $I(t) = \mu(t) = N_C \times i \times P_1$, where is $I(t)$ is the ensemble average current and P_1 represents the open probability.

The autocovariance is given by:

$$C(t_1, t_2) = N_c i^2 \{P_1(t_1)P_{11}(t_2 \mid t_1) - P_1(t_1)P_1(t_2)\}$$

where $P_{11}(t_2|t_1)$ is the conditional probability that the channel is open at t_2 given that it was open at t_1.

The variance is given by:

$$\sigma^2(t_1) = C(t_1, t_1) = N_c i^2 \{P_1(t_1) \times 1 - P_1(t_1)^2\}$$
$$= N_c i^2 \{P_1(t_1)(1 - P_1(t_1))\}$$

This relationship can be written in terms of the mean current as follows (Sigworth 1980):

$$\sigma^2(t_1) = i \times I(t) - \frac{I(t)^2}{N_C}$$

Plotting the variance as a function of the mean current yields a parabola, which provides a convenient way to determine single-channel properties from macroscopic measurements. Specifically, the initial slope corresponds to the unitary current amplitude. The intercepts, where the variance is zero, are attained when $P_1 = 0$, or when all channels are closed, or when $P_1 = 1$ when all channels are open. The maximal variance occurs when $P_1 = 0.5$ (figure 5.1).

A similar approach can be used to study fluctuations of nonstationary current (Sigworth 1981). The key difference is that statistical parameters including $\mu(t)$ and $\sigma^2(t)$ are obtained experimentally by analyzing an ensemble of records and cannot be determined from a single long-term record. This approach may be particularly advantageous when studying voltage-gated ion channels, as it allows one to establish the number of channels and other elementary properties of ion channels (Alvarez et al 2002).

A compelling example of the use of this method to gain biological insight comes from classic work considering the mechanism of adrenergic regulation of L-type channels (Bean et al 1984). The sympathetic nervous system enhances L-type calcium currents and boosts contractility in ventricular cardiomyocytes. In this study, fluctuation analysis of whole-cell calcium currents from frog myocytes was used to compare changes in unitary current, open probability, and the number of channels following adrenergic stimulus (figure 5.2) (Bean et al 1984). The unexpected finding here was that adrenergic stimulus caused a marked increase in the number of 'functional channels' at the membrane, along with a small increase in open probability (P_O) and a negligible change in unitary current. This result

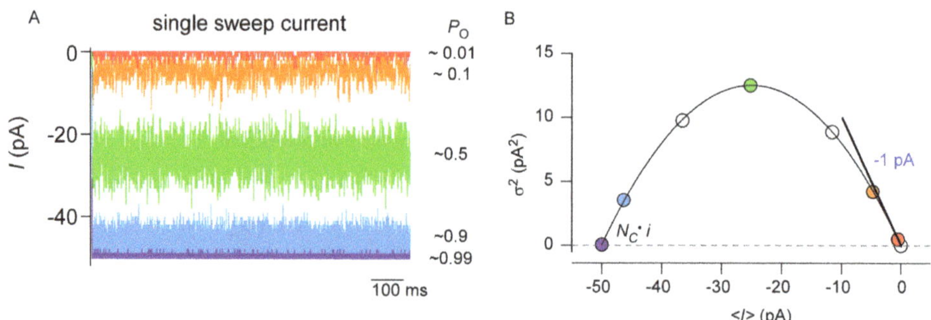

Figure 5.1. Mean-variance analysis allows for the estimation of the single-channel parameters. (A) Stochastic simulation of a two-state ion channel with varying P_O. The variance of the fluctuation in the current trace is small when $P_O\sim0$ or when $P_O\sim1$. The highest variability is observed at $P_O\sim0.5$. (B) Parabolic dependence of $\sigma^2(t_1)$ on the mean current amplitude $I(t)$. The initial slope yields the unitary current amplitude (i), and the intercept is $\sim N_C \cdot i$.

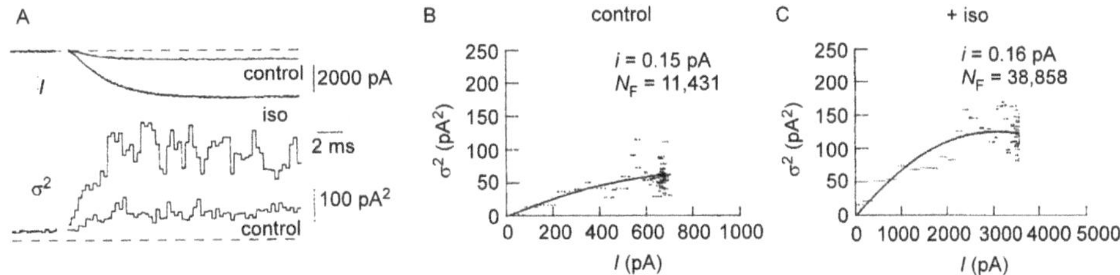

Figure 5.2. Mean-variance analysis of calcium current in frog ventricular cardiomyocytes to identify changes in channel gating following adrenergic stimulation. (A) Mean current (top) and the baseline variance of calcium current (control) and with isoproterenol. (B and C) Mean variance analysis shows that isoproterenol increases the number of functional channels available to open. Parts (A)–(C) reproduced from [Bean *et al* 1984], with permission from Springer Nature.

suggested that at rest, a fraction of the calcium channels remained silent for prolonged periods (i.e. longer than the interpulse interval) and only underwent occasional transitions to a functional configuration where channels readily opened within milliseconds. This represents a pool of channels that form a 'functional reserve.' Following phosphorylation of the ion-channel complex during the fight-or-flight response, the previously silent channels become available for rapid gating. It is important to note here that the number of functional channels is distinguished from the total number of channels in the membrane, which includes both the rapidly gating channels and the silent ones. These results predicted distinct 'gating modes' of calcium channels, which were later substantiated by single-channel recordings.

5.4 Stationary ergodic systems

Ergodicity is a property of a stochastic process in which the statistical information contained in a single sufficiently long realization is equivalent to the information obtained from an ensemble of many independent realizations (figure 5.3). Thus, ergodicity bridges time averages (from one long record) with ensemble averages (across many realizations). From a practical perspective, this means that for a stationary ergodic process, we can obtain all the results we need from any single record, given that the record is sufficiently long.

Mathematically, for any process, the mean is:

$$\mu(t) = \mu = <x_j(t)>_t = <x_i(t)>_i; \text{ for all } j \text{ and } t,$$

where $\langle x_j(t)\rangle_t$ represents a time average, while $\langle x_i(t)\rangle_i$ represents the ensemble average (figure 5.3B).

The autocovariance for a small lag τ is given by (figure 5.3C):

$$C(t, t + \tau) = C(\tau)$$
$$= <(x_j(t) - \mu)(x_j(t + \tau) - \mu)>_t$$
$$= <(x_i(t) - \mu)(x_i(t + \tau) - \mu)>_i \text{ for all } i \text{ and } t.$$

5.5 Stationary ergodic systems lend themselves to frequency-domain tools

From a broader perspective, single-channel kinetics can be modeled as a Markov process governed by a transition rate matrix Q. The matrix exponential $\exp(Q{\cdot}t)$ fully specifies the probability of the system occupying a given state as a function of time and thus fully characterizes the underlying gating dynamics.

As noted above, the kinetic description lends itself to the time-domain tools discussed above. The key parameter here is the autocovariance function $C(t_1, t_2)$, which measures how fluctuations in channel activity at time t_2 are related to those at a previous time t_1. For stationary ergodic systems, the Wiener–Khinchin theorem relates the autocovariance function to the frequency domain using the power spectral density function (PSDF). This equivalence allows the characteristic time-scales of ion-channel gating to be obtained either from time-domain correlation analysis or from frequency-domain spectral analysis (figure 5.4).

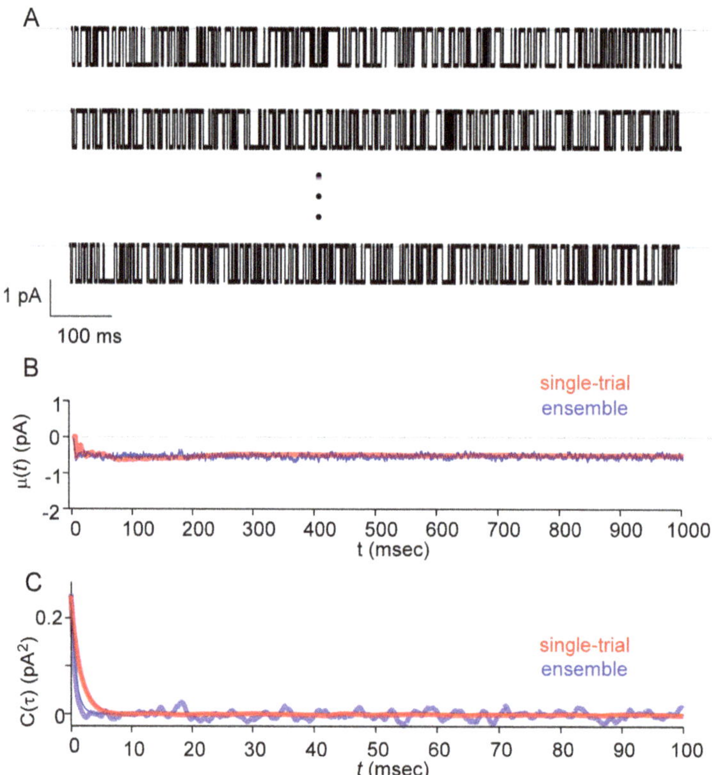

Figure 5.3. Simulation illustrating the concept of ergodicity. (A) Three sample functions or records from a stochastic simulation of a two-state ion channel. (B) The single-trial average current (red) approximates the ensemble average (blue) calculated across different records. The single-trial average current is calculated here as a moving average. (C) Autocovariance obtained from a single record (red) and averaged across many records (blue).

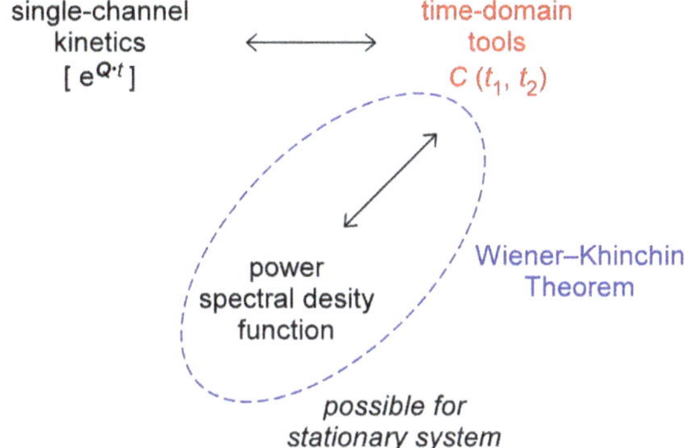

Figure 5.4. Schematic showing the interrelationship between single-channel kinetics and time-domain tools. For stationary ergodic processes, the Wiener–Khinchin theorem allows the time-domain tools to be extended into the frequency domain via the PSDF.

The PSDF describes how the variance or power of a signal is distributed in the frequency spectrum. Intuitively, it describes which frequencies contribute most to the fluctuations in the signal. For example, 'white noise' has equal power spectral density across all frequencies.

For a zero-mean stationary signal truncated to a finite observation window,

$$\Delta I_T(t) = \begin{cases} I(t) - \mu & -T \leqslant t \leqslant T \\ 0 & \text{otherwise} \end{cases}.$$

The power spectral density is given by:

$$S_{\Delta I}(f) \equiv \text{PSDF} \equiv \lim_{T \to \infty} \frac{1}{2T} |\Delta I_T(f)|^2,$$

where $\Delta I_T(f)$ is the Fourier transform of $\Delta I_T(t)$.

The Wiener–Khinchin theorem states that

$$S_{\Delta I}(f) = \int_{t=-\infty}^{+\infty} C(\tau) \cdot e^{-j2\pi f \tau} d\tau.$$

For a stationary system, the autocovariance is an even function, meaning that $C(\tau) = C(-\tau)$. Conceptually, as the statistical properties of a stationary process does not change, the covariance of a signal at the present time and the same signal a few milliseconds later is the same as the covariance between the signal at the present time and the signal a few milliseconds earlier. More formally, this equivalence can be obtained as follows:

$$C(-\tau) = \langle \Delta I(t)\Delta I(t - \tau)\rangle_{t:-\infty \to +\infty} = \langle \Delta I(t - \tau)\Delta I(t)\rangle_{t:-\infty \to +\infty}$$
$$= \langle \Delta I(\hat{t})\Delta I(\hat{t} + \tau)\rangle_{\hat{t}:-\infty \to +\infty} = C(+\tau).$$

This implies that that $S_{\Delta I}(f) = S_{\Delta I}(-f)$. In this case, it is often convenient to define a one-sided PSDF, $G_{\Delta I}(f)$:

$$G_{\Delta I}(f) = \begin{cases} S_{\Delta I}(0), & f = 0 \\ 2S_{\Delta I}(f), & f > 0. \\ N.\,D., & f < 0 \end{cases}$$

From a practical perspective, the covariance of a Markov process is often an exponential function. The Fourier transform of an exponential function yields the Lorentzian below:

$$\mathcal{F}\{e^{-a|t|}\} = \frac{2a}{a^2 + (2\pi f)^2}.$$

Proof of the Wiener–Khinchin theorem

For $\Delta I_T(t)$ as defined above, its Fourier transform can be obtained as follows:

$$\Delta I_T(f) \equiv \int_{-\infty}^{+\infty} \Delta I_T(t)e^{-j2\pi ft}dt.$$

Let

$$C_T(\tau) = \frac{1}{2T}\int_{-\infty}^{+\infty} \Delta I_T(t)\Delta I_T(t+\tau)dt \cong \frac{1}{2T}\int_{-T}^{T} \Delta I_T(t)\Delta I_T(t+\tau)d\tau,$$

such that

$$C_T(\tau) \equiv \lim_{T \to \infty} C_T(\tau).$$

The limit exists if the process is stationary.

In this case,

$$C_T(f) \equiv \int_{\tau=-\infty}^{\infty} C_T(\tau)e^{-j2\pi f\tau}d\tau = \int_{\tau=-\infty}^{\infty} \frac{1}{2T}\left[\int_{t=-\infty}^{+\infty} \Delta I_T(t)\Delta I_T(t+\tau)dt\right]e^{-j2\pi f\tau}\,d\tau$$

$$= \frac{1}{2T}\int_{\tau=-\infty}^{\infty}\int_{t=-\infty}^{\infty} (\Delta I_T(t)e^{+j2\pi ft})(\Delta I_T(t+\tau)e^{-j2\pi f(t+\tau)})dtd\tau.$$

Switching the order of the integration yields

$$= \frac{1}{2T}\int_{t=-\infty}^{\infty}\left[\int_{\tau=-\infty}^{\infty} \Delta I_T(t+\tau)e^{-j2\pi f(t+\tau)}d\tau\right]\Delta I_T(t)e^{+j2\pi ft}dt.$$

We apply a change of variables $u = t + \tau$ with t fixed and $du = d\tau$:

$$= \frac{1}{2T}\int_{t=-\infty}^{\infty}\left[\int_{u=-\infty}^{\infty} \Delta I_T(u)e^{-j2\pi fu}du\right]\Delta I_T(t)e^{+j2\pi ft}dt.$$

Recognizing that the inner integral is $\Delta I_T(f)$, which is independent of t, the expression is simplified to

$$= \frac{1}{2T}\Delta I_T(f)\int_{t=-\infty}^{\infty} \Delta I_T(t)e^{+j2\pi ft}dt.$$

The latter term is the complex conjugate of $\Delta I_T(f)$. Thus,

$$C_T(f) = \frac{1}{2T}\Delta I_T(f)\overline{\Delta I_T(f)} = \frac{1}{2T}|\Delta I_T(f)|^2.$$

This leads to a series of equivalencies:

$$C_T(f) = \mathcal{F}\{C(\tau)\} = \mathcal{F}\left\{\lim_{T\to\infty} C_T(\tau)\right\} = \int_{\tau=-\infty}^{\infty} e^{-j2\pi f\tau}\left(\lim_{T\to\infty} C_T(\tau)\right)d\tau$$

$$= \lim_{T\to\infty}\int_{\tau=-\infty}^{\infty} e^{-j2\pi f\tau}\,C_T(\tau)d\tau = =\lim_{T\to\infty}\frac{1}{2T}|\Delta I_T(f)|^2 = S_{\Delta I}(f).$$

5.6 Example: a simple two-state channel

To illustrate how the frequency-domain tools relate to single-channel gating parameters from fluctuation analysis, consider a simplified two-state channel with a closed nonconductive state and an open conductive state,

$$C \underset{\beta}{\overset{\alpha}{\rightleftharpoons}} O.$$

Based on the analysis in chapter 4,

$$\frac{dP_O}{dt} = \alpha P_C - \beta P_O = \alpha - (\alpha + \beta)P_O$$

In the steady state, $P_O(t)$ is

$$P_O(\infty) = \frac{\alpha}{\alpha + \beta}.$$

For this model,

$$P_{OO}(t_1 + \tau \,|\, t_1) = P_{OO}(\tau\,|\,0) = \frac{\alpha}{\alpha + \beta} + \frac{\beta}{\alpha + \beta}e^{-(\alpha+\beta)t}.$$

The covariance

$$C(t_1, t_2) = N_c i^2 P_O(t_1)\{P_{OO}(t_2 \,|\, t_1) - P_O(t_2)\},$$

where N_C is the number of channels and i is unitary current amplitude. Assuming the recordings are obtained in the steady state to satisfy the stationary condition,

$$C(\tau) = N_c i^2 P_O(\infty)\{P_{OO}(\tau\,|\,0) - P_O(\infty)\}$$

$$= N_c i^2 P_O(\infty)\left\{\frac{\alpha}{\alpha + \beta} + \frac{\beta}{\alpha + \beta}e^{-(\alpha+\beta)t} - \frac{\alpha}{\alpha + \beta}\right\}$$

$$= N_c i^2 P_O(\infty) \times (1 - P_O(\infty))e^{-(\alpha+\beta)\tau}.$$

The PSDF is given by

$$S_{\Delta I}(f) = \frac{2N_c i^2 P_O(\infty) \times (1 - P_O(\infty))(\alpha + \beta)}{(\alpha + \beta)^2 + (2\pi f)^2}.$$

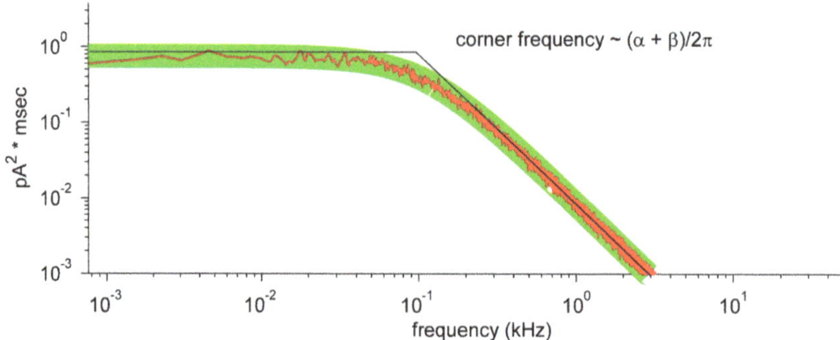

Figure 5.5. PSDF for a simulated two-state ion channel. The corner frequency is related to single-channel kinetics. Below the corner frequency, the PSDF is flat. At higher frequencies, the PSDF falls by $\sim 1/f^2$ (40 decibels per decade).

In terms of the mean current $\mu = N_c i P_O(\infty)$,

$$S_{\Delta I}(f) = \frac{2i \times \mu \times (1 - P_O(\infty))(\alpha + \beta)}{(\alpha + \beta)^2 + (2\pi f)^2}.$$

Graphically, the PSDF is flat for frequencies lower than the corner frequency but decays for higher frequencies (figure 5.5).

5.7 Example: neurotransmitter binding to ligand-gated ion channels

At the neuromuscular junction, the release of acetylcholine (ACh) activates nicotinic acetylcholine receptors (nAChRs) in the postsynaptic membrane (Anderson and Stevens 1973). The stochastic opening and closing of these ligand-gated ion channels generates fluctuations in the end-plate conductance. Analysis of these fluctuations provides insight into the microscopic kinetics of the receptor, including transition rates, even when individual channel openings are not directly resolved.

Consider the following model for the nAChR channel, which binds n neurotransmitters (T) to activate. The concentration of ACh is given by C with an association constant K:

$$R \underset{}{\overset{KC^n}{\rightleftharpoons}} T_n R \underset{\alpha}{\overset{\beta}{\rightleftharpoons}} T_n R^*.$$

At low concentrations of ACh, the probability of the unbound state is near unity $(P_R \sim 1)$, and the occupancy of the bound closed state can be approximated by $P_{T_n R} = KC^n \times R \approx KC^n$.

If $P_O(t)$ denotes the open probability of the channel, then its time evolution depends on

$$\frac{dP_O(t)}{dt} = -(\alpha)P_O(t) + \beta \times K \times C^n.$$

In the steady state, the open probability $P_0(\infty)$ is given by

$$P_0(\infty) = \frac{\beta K C^n}{\alpha}.$$

The conditional open probability $P_{OO}(\tau)$ is given by

$$P_{OO}(\tau) = \frac{\beta k c^n}{\alpha} + \left(1 - \frac{\beta k c^n}{\alpha}\right)e^{-\alpha\tau}.$$

Assuming a unitary current amplitude of i for the open channel, the autocovariance for a single channel can be calculated as follows:

$$C(\tau) = i^2 P_0(\infty)(P_{OO}(\tau) - P_0(\infty)).$$

Substituting $P_0(\infty)$ and $P_{OO}(\tau)$ yields

$$= i^2\frac{\beta K C^n}{\alpha}\left(\frac{\beta K C^n}{\alpha} + \left(1 - \frac{\beta K C^n}{\alpha}\right)e^{-\alpha\tau} - \frac{\beta K C^n}{\alpha}\right)$$

$$= i^2\frac{\beta K C^n}{\alpha}\left(1 - \frac{\beta K C^n}{\alpha}\right)e^{-\alpha\tau}.$$

At low concentrations of ACh, $\frac{\beta K C^n}{\alpha} \ll 1$; therefore,

$$C(\tau) \cong i^2\frac{\beta K C^n}{\alpha}e^{-\alpha\tau}.$$

For N independent and identical channels,

$$C_N(\tau) \approx \frac{Ni^2\beta K C^n}{\alpha}e^{-\alpha\tau}.$$

The Fourier transform of the exponential autocovariance yields a Lorentzian PSDF:

$$S_{\Delta I}(f) = \frac{Ni^2\beta K C^n}{\alpha}\frac{2\alpha}{\alpha^2 + (2\pi f)^2} = \frac{Ni^2\beta K C^n/\alpha^2}{1 + (2\pi f/\alpha)^2}.$$

Since the mean current is given by

$$\mu_I \approx Ni\frac{\beta K C^n}{\alpha},$$

the PSDF can be denoted by:

$$S_{\Delta I}(f) = \frac{2i\mu_I/\alpha}{1 + (2\pi f/\alpha)^2}.$$

In this case, the corner frequency is given by $\alpha/2\pi$. Below this frequency, the PSDF is relatively flat, but at higher frequencies, it falls by $\sim 1/f^2$ (i.e. with a slope of ~ 2 on a log–log axis). These studies provided the first insights into the elementary properties of the nAChR (figure 5.6).

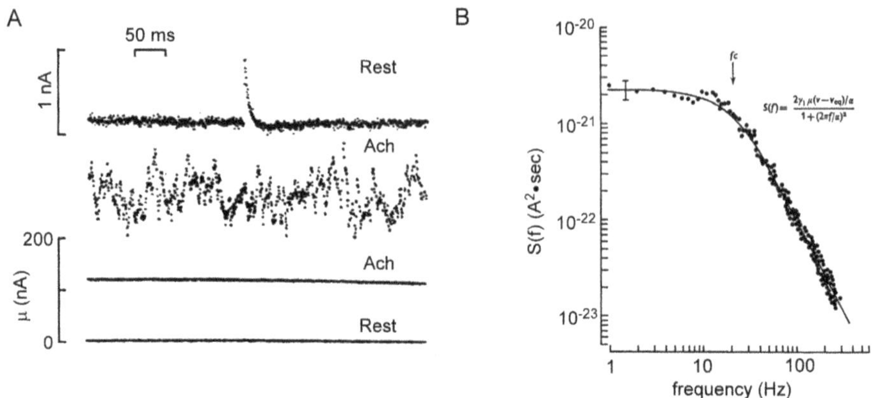

Figure 5.6. Spectral analyses of fluctuations in muscle end-plate currents produced by the constant application of ACh. (A) Membrane currents in a muscle end plate under a voltage clamp. (B) Power spectral density of fluctuations produced by constant ACh application. The corner frequency, $f_c = \alpha/(2\pi) \sim 21$ Hz. Parts (A) and (B) [Anderson & Stevens, 1973] John Wiley & Sons. © 1973 The Physiological Society.

References

Alvarez O, Gonzalez C and Latorre R 2002 Counting channels: a tutorial guide on ion channel fluctuation analysis *Adv. Physiol. Educ.* **26** 327–41

Anderson C R and Stevens C F 1973 Voltage clamp analysis of acetylcholine produced end-plate current fluctuations at frog neuromuscular junction *J. Physiol.* **235** 655–91

Bean B P, Nowycky M C and Tsien R W 1984 Beta-adrenergic modulation of calcium channels in frog ventricular heart cells *Nature* **307** 371–5

Sigworth F J 1980 The variance of sodium current fluctuations at the node of Ranvier *J. Physiol.* **307** 97–129

Sigworth F J 1981 Covariance of nonstationary sodium current fluctuations at the node of Ranvier *Biophys. J.* **34** 111–33

IOP Publishing

Ion Channel Gating and Mechanisms

David T Yue, Manu Ben-Johny and Ivy E Dick

Chapter 6

Gating currents: an alternate view of channel operation

6.1 Basic notion of gating currents

A voltage-gated channel must have charged residues that 'move' in response to the transmembrane electric field. This is the irreducible physical basis of a 'voltage sensor,' the movement of which triggers conformational changes leading to channel opening. It turns out that the intramembrane movement of these charged residues is inextricably linked to reciprocal movements of charge outside of the membrane. These 'gating-charge' movements are measurable in the absence of ionic current through the pore of the channels. In contrast to ionic currents (which emanate from ionic flux through open states only), such 'gating currents' provide a signature of the channel as it moves through closed conformations. In this manner, gating currents and gating-charge redistribution provide data that furnish a different and usually stronger reflection of closed-state behavior than standard ionic current measurements (Hodgkin and Huxley 1952, Armstrong and Bezanilla 1973, Colquhoun and Hawkes 1981, Catacuzzeno *et al* 2023).

6.2 Physical basis

Consider the cell membrane in which channels reside as an infinite flat sheet of insulation separating two infinite conductors. This is the simplest model of a parallel-plate capacitor. By symmetry, the electric field lines must be perpendicular to the sheets of free surface charge, with density $\pm\sigma_f$ (figure 6.1). By the same token, the electric field strength must be independent of the distance from the sheets of charge. Hence, the field in either the right or the left conductor must be zero because it reflects the net influence of positive and negative sheets of charge with equal magnitudes of surface charge density. A very useful electrostatic law in this context is Gauss's law:

$$\oint \vec{E} \times \overrightarrow{dA} = 4\pi kQ.$$

doi:10.1088/978-0-7503-2388-8ch6

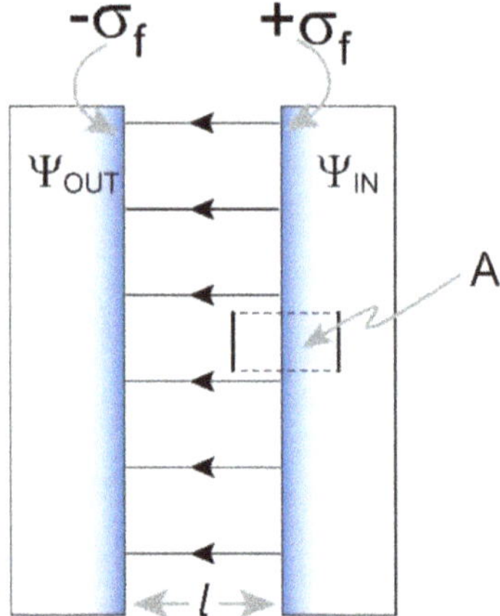

Figure 6.1. Electrostatic model of the cell membrane as a parallel-plate capacitor. The membrane acts as an insulating layer between two conductive regions with surface charge densities $\pm\sigma_f$. Electric field lines are perpendicular to the plates, and the field strength is uniform across the insulator.

A box drawn as shown below can be used to calculate the strength of the electric field in the insulator. Applying Gauss's law yields

$$E \times A = 4\pi k A \sigma_f$$

or $E = 4\pi k \sigma_f$. The voltage drop across the insulator is therefore $V_m = 4\pi k \sigma_f l$. The capacitance C of the membrane is given by Q/V, which yields:

$$C \equiv \frac{Q}{V} = \frac{\sigma_f A}{4\pi k \sigma_f l} = \frac{A}{4\pi k l}.$$

Now, what happens when we place a charged residue in the insulator (figure 6.2), say to mimic a portion of a channel's voltage sensor? Under a voltage clamp at a fixed transmembrane voltage drop of V, there is a unique distribution of reciprocal 'neutralization charges' that must result when such a charged residue is inserted into the membrane. Let us say the charge is q. The charge is placed at a distance $f_p l$ from the left edge of the insulator, where f_p ranges from zero to one. The sum of the neutralization charges on the right and left edges of the insulator, $q_{neutral,L} + q_{neutral,R}$, can be calculated by a very simple argument. The electric field in a conductor must be zero, by definition of a conductor. In either the right or left conductor, at a large distance from the charge q, the intramembrane charge and neutralization charges appear to be a single point charge with a net charge equal to $q + q_{neutral,L} + q_{neutral,R}$. For the electric field in the conductor to be zero, this sum must be zero. Hence,

$$-q = q_{neutral,L} + q_{neutral,R}.$$

To determine the exact values of the right and left neutralization charges, we have to digress momentarily to an alternate idealized model. Consider a sheet of charge in

$$-q = q_{neutral,L} + q_{neutral,R}.$$

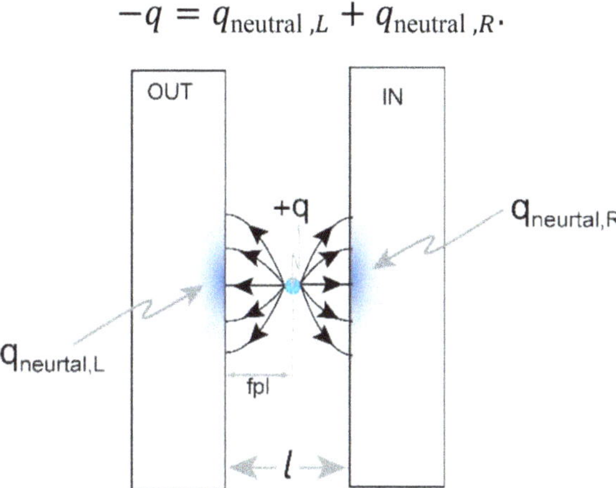

Figure 6.2. Introduction of a point charge into the membrane, showing the effect of inserting a charged residue (q) into the membrane at a fractional position $f_p l$.

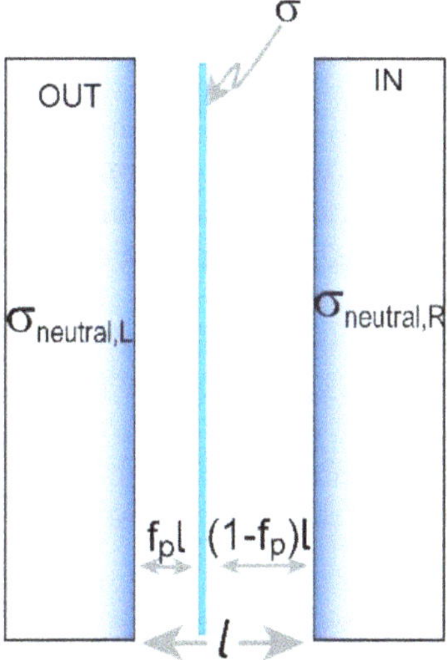

Figure 6.3. Sheet charge model of neutralization. Idealized model showing a sheet of charge embedded in the membrane and the resulting neutralization charge densities on either side.

the membrane with surface charge density σ, arrayed at a distance $f_p l$ from the left edge of the insulator (figure 6.3).

From Gauss's law, it is easy to establish that the electric field emanating from a sheet of charge (say the intramembranous sheet) has a magnitude of $2\pi k \sigma$. For the field in the left and right conductors to be zero, the sum of the left and right neutralizing sheets must be as follows:

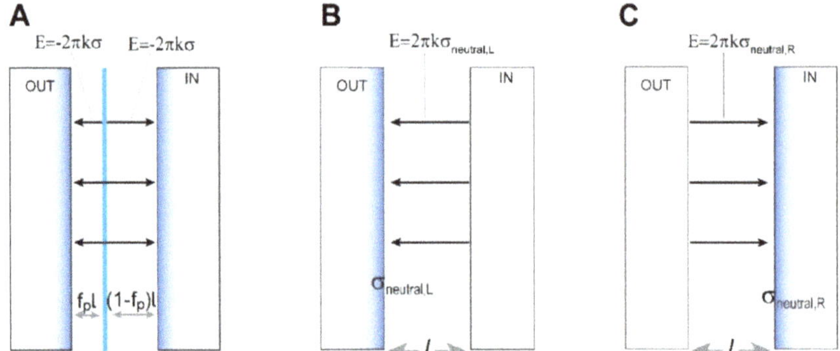

Figure 6.4. Voltage contributions from different charge distributions. (A) Voltage change due to the intramembranous charge density, proportional to its position. (B) Voltage change produced by the left neutralization charge density. (C) Voltage change produced by the right neutralization charge density.

$$\sigma_{\text{neutral}, L} + \sigma_{\text{neutral}, R} = -\sigma.$$

A second constraint comes from the fact that the addition of neutralization and intramembranous charge sheets should not perturb the transmembrane voltage from what it was before this addition, as we are under voltage-clamp conditions. To ensure that no such perturbation occurs, the neutralization charge densities must also be arranged so that the net voltage drop across the membrane—the drop caused by the intramembranous and neutralization charge densities—remains zero (figure 6.4).

The net voltage change (inside (right) potential minus outside (left) potential) produced by the intramembranous charge density is given by $2\pi k\sigma\,(f_p - (1 - f_p)) \times l$ (basically the electric field times the distance; see panel A). The net voltage change (inside (right) potential minus outside (left) potential) produced by the left neutralization charge density is given by $-2\pi k\sigma_{\text{neutral},L} \times l$ (see panel B). The net voltage change (inside (right) potential minus outside (left) potential) produced by the right neutralization charge density is given by $2\pi k\sigma_{\text{neutral},R} \times l$ (see panel C). Adding these voltage changes together yields a second constraint,

$$\sigma(2f_p - 1) - \sigma_{\text{neutral}, L} + \sigma_{\text{neutral}, R} = 0.$$

Solving these two constraints simultaneously yields the satisfyingly simple result that

$$\sigma_{\text{neutral}, L} = -(1 - f)\sigma \quad \text{and} \quad \sigma_{\text{neutral}, R} = -f\sigma.$$

The smooth surface charge density σ can be conceptualized as the net behavior of many smoothly distributed, equal-magnitude, infinitesimally small point charges. By symmetry, the contribution of each such point charge (say with charge q) to the neutralization charge must be the same. Hence, $q_{\text{neutral},L} \propto -(1 - f) \times q$ and $q_{\text{neutral},R} \propto -f \times q$. We already know from our point charge model above that $-q = q_{\text{neutral},L} + q_{\text{neutral},R}$. Hence, the proportionalities can be replaced by equalities. Because what holds for an infinitesimal point charge must hold for an arbitrary

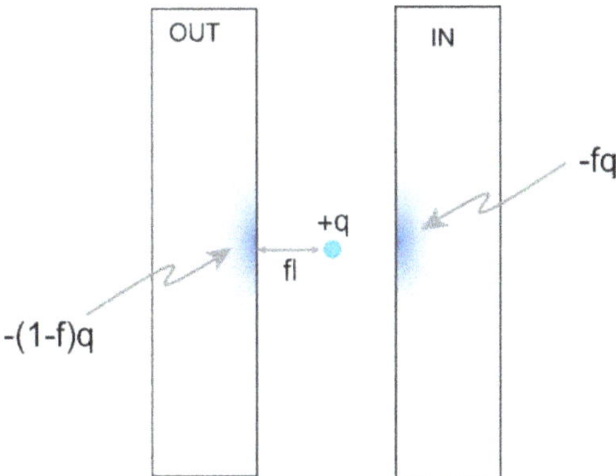

Figure 6.5. Model of gating charge and neutralization. Diagram summarizing the effect of a point charge q in the membrane, showing the resulting neutralization charges on the left and right surfaces. The position f of the charge varies with the channel state, leading to the dynamic rearrangement of neutralization charges and the generation of gating currents.

point charge in terms of the behavior of the neutralization charges, the following must be true of the neutralization charges for the original model following the insertion into the membrane of an arbitrary point charge q (figure 6.5):

$$q_{\text{neutral},L} = -(1 - f) \times q$$

$$q_{\text{neutral}, R} = -f \times q.$$

The final solution for the effect of a point charge q in the membrane is diagrammed below. The position f of the charge changes in the different states of the channel. Hence, rearrangements of neutralization charges occur as the channel moves between different states, leading to gating currents. The actual case is only complicated by the fact that there are an arbitrary number m of charged residues in the channel. Since the effects of charges just add according to the superposition principle, gating currents arise in an analogous fashion in the general case.

6.3 Effect of gating-charge movement on the Q–V curve

Now consider a channel with m intramembranous charged residues, each with charge q_i. Suppose that there are n states of the channel. Suppose that the position of the ith charged residue in state j is given by f_{ij}; then, the neutralization charge on the right conductor that results in state j is given by

$$q_{\text{neutral}, R, j} = -(f_{1j} q_1 + \cdots + f_{mj} q_m).$$

From the standpoint of actual measurements, we can never distinguish the absolute neutralization charge magnitudes on the right and left conducting plates; we can only assess differences in neutralization charge as the channel changes between states. For convenience, we take as our reference point the arbitrary condition when the channel is in state 1, where the channel resides with a probability of unity at infinite hyperpolarization. In this condition, we consider there to be no neutralization charge on the right and left plates. Moving to other states then results in a redistribution of neutralization charges, leading to the deposition of q on the right plate and the deposition of $-q$ on the left plate. With this insight, we can refer to the measurable charge q as the channel changes from state 1 to state j as equalling $q_{\text{neutral},R,j} - q_{\text{neutral},R,1}$ above. We will henceforth refer to it simply as Δq_{1j}, given by

$$\Delta q_{1j,R} = q_{\text{neutral},R,j} - q_{\text{neutral},R,1}$$
$$= -[(f_{1j} - f_{11})q_1 + \cdots + (f_{mj} - f_{m1})q_m].$$

The neutralization charge for the channel $q_{\text{neutral,channel}}$ is therefore a random variable that can adopt the different values of q_{1j}. The expectation value of $\Delta q_{\text{neutral, channel}}$ is therefore given by

$$\left\langle q_{\text{neutral, channel}} \right\rangle = \sum_{j \in \text{ all states}} \Delta q_{1j} P_j(t)$$

$$= \begin{bmatrix} P_1(t) & \cdots & P_n(t) \end{bmatrix} \begin{bmatrix} \Delta q_{11} = 0 \\ \Delta q_{12} \\ \vdots \\ \Delta q_{1n} \end{bmatrix}$$

$$= \begin{bmatrix} P_1(0) & \cdots & P_n(0) \end{bmatrix} [e^{Qt}] \begin{bmatrix} \Delta q_{11} = 0 \\ \Delta q_{12} \\ \vdots \\ \Delta q_{1n} \end{bmatrix}$$

If the state names are chosen such that states with increasing index values are successively favored at increasing depolarization, then we can plot the steady state $\left\langle q_{\text{neutral, channel}} \right\rangle$ as a function of voltage, along with the steady state $P_i(\infty)$ as a function of voltage, as shown in the top two panels below. The $Q_C - V$ relation for the membrane capacitance is given in the third panel. The total steady-state $Q_{\text{tot}} - V$ curve is the sum of the steady state $\left\langle q_{\text{neutral, channel}} \right\rangle + Q_C$, as shown in the bottom panel. By performing leak subtraction over a negative voltage range, $\left\langle q_{\text{neutral, channel}} \right\rangle$ alone can be extracted experimentally (figure 6.6).

The interesting philosophical point is that gating-charge movement thus gives us a different 'lens' through which to view the workings of the channel $[e^{Qt}]$. In the case of macroscopic ionic currents, the lens was a column vector with zeros everywhere except the open state(s). For example, for a single open state n:

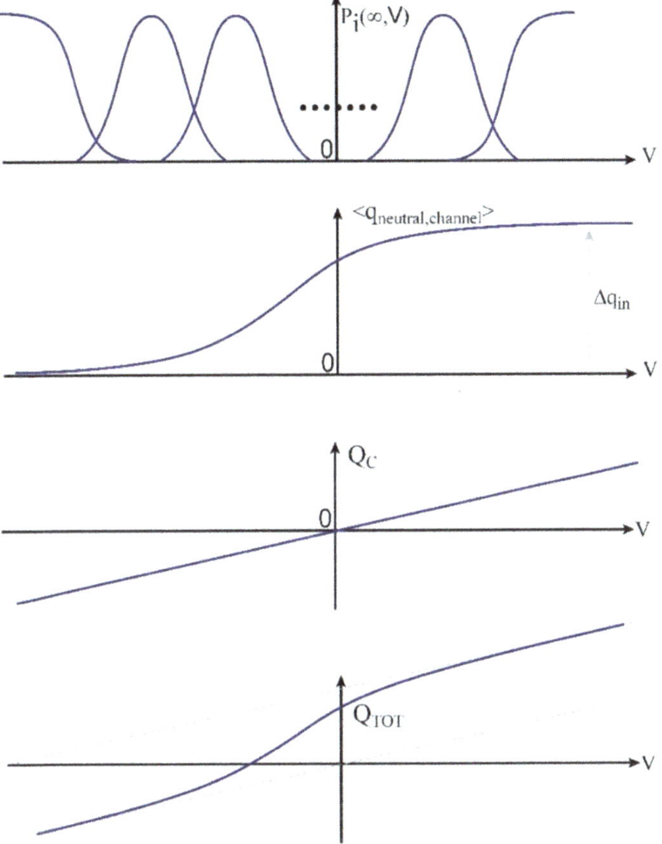

Figure 6.6. Voltage dependence of gating charge and capacitance.

$$I = \mathrm{NiP_{open}}(t) = \mathrm{NiP}_a(0)[e^{Qt}]\begin{bmatrix} 0 \\ \vdots \\ 0 \\ 1 \end{bmatrix}.$$

For duration histograms, we always placed the interface between transient and absorbing classes at the boundary separating open and other states. We then looked through the lens of $[Q_{TA}]$, subject to such a margin. Recall

$$[f_{ij}(t)] = \mathbf{P}_a(0)[e^{Qt}][Q_{TA}].$$

Now, however, we look through the lens of

$$\begin{bmatrix} \Delta q_{11} = 0 \\ \Delta q_{12} \\ \vdots \\ \Delta q_{1n} \end{bmatrix},$$

which 'weights' deep closed states considerably more than the earlier two lenses.

6.4 Gating-current formalism

To determine the gating current, we are interested in

$$<i_{\text{gate}}(t)> = \frac{d\langle q_{\text{neutral, channel}}\rangle}{dt}$$

$$= \sum_{j \in \text{ all states}} \Delta q_{1j} \frac{dP_j(t)}{dt}$$

$$= \left[\frac{dP_1(t)}{dt} \cdots \frac{dP_n(t)}{dt} \right] \begin{bmatrix} \Delta q_{11} = 0 \\ \Delta q_{12} \\ \vdots \\ \Delta q_{1n} \end{bmatrix} \tag{6.1}$$

$$= [P_1(0) \cdots P_n(0)][e^{Qt}][Q] \begin{bmatrix} \Delta q_{11} = 0 \\ \Delta q_{12} \\ \vdots \\ \Delta q_{1n} \end{bmatrix},$$

which gives us the lens

$$[Q] \begin{bmatrix} \Delta q_{11} = 0 \\ \Delta q_{12} \\ \vdots \\ \Delta q_{1n} \end{bmatrix}.$$

Example 1 *Consider the simple two-state gating scheme below, with the gating-charge parameter q_{12}:*

$$C_1 \underset{k_{21}}{\overset{k_{12}}{\rightleftharpoons}} O_2.$$

Putting this into the formalism for gating current, we get

$$<i_{\text{gate}}(t)> = [P_1(0)\ P_2(0)][e^{Qt}] \begin{bmatrix} -k_{12} & k_{12} \\ k_{21} & -k_{21} \end{bmatrix} \begin{bmatrix} \Delta q_{11} = 0 \\ \Delta q_{12} \end{bmatrix}$$

$$= \underbrace{P_1(t)k_{12}}_{\substack{\text{average number of} \\ 1 \to 2 \text{ transitions} \\ \text{per unit time} \\ \text{per channel}}} \times \underbrace{\Delta q_{12}}_{\substack{\text{charge per} \\ 1 \to 2 \text{ transition}}}$$

$$+ \underbrace{P_2(t)k_{21}}_{\substack{\text{average number of} \\ 2 \to 1 \text{ transitions} \\ \text{per unit time} \\ \text{per channel}}} \times \underbrace{(-\Delta q_{12})}_{\substack{\text{charge per} \\ 2 \to 1 \text{ transition}}},$$

which gives an intuitive feel for how to form the average gating current. This approach builds on the stochastic framework developed by Colquhoun and Hawkes (1981), which remains foundational in ion-channel modeling.

References

Armstrong C M and Bezanilla F 1973 Currents related to movement of the gating particles of the sodium channels *Nature* **242** 459–61

Catacuzzeno L, Conti F and Franciolini F 2023 Fifty years of gating currents and channel gating *J. Gen. Physiol.* **155** e202313380

Colquhoun D and Hawkes A G 1981 On the stochastic properties of single ion channels *Proc. R. Soc. Lond. B: Biol. Sci.* **211** 205–35

Hodgkin A L and Huxley A F 1952 A quantitative description of membrane current and its application to conduction and excitation in nerve *J. Physiol.* **117** 500–44